A Contract with the Earth

To
Peter —
Best wishes
Newt [signature]

A
Contract
with the
Earth

NEWT GINGRICH & TERRY L. MAPLE

Foreword by E. O. Wilson

THE JOHNS HOPKINS UNIVERSITY PRESS

Baltimore

The Johns Hopkins University Press
2715 North Charles Street
Baltimore, Maryland 21218-4363
www.press.jhu.edu

Library of Congress Cataloging-in-Publication Data
Gingrich, Newt.
A contract with the Earth / Newt Gingrich & Terry L. Maple ;
foreword by E. O. Wilson.
p. cm.
Includes bibliographical references and index.
ISBN-13: 978-0-8018-8780-2 (hardcover : acid-free paper)
ISBN-10: 0-8018-8780-1 (hardcover : acid-free paper)
1. Environmental responsibility—United States.
2. Environmentalism—United States. 3. Environmental policy—
United States. I. Maple, Terry. II. Title.
GE195.7.G56 2007
333.720973—dc22   2007020431

A catalog record for this book is available from the British Library.

*Special discounts are available for bulk purchases of this book.
For more information, please contact Special Sales at 410-516-6936
or specialsales@press.jhu.edu.*

The Johns Hopkins University Press uses environmentally friendly
book materials, including recycled text paper that is composed of
at least 30 percent post-consumer waste, whenever possible.
All of our book papers are acid-free, and our jackets and covers are
printed on paper with recycled content.

We dedicate

A Contract with the Earth *to our wives,*
Callista and Addie, our daughters, Jackie, Kathy,
Molly, Emily, and Sally, and our grandchildren,
who will surely enjoy a lifetime of peace and
prosperity on a cleaner, greener, and thoroughly
renewed Earth.

CONTENTS

FOREWORD

A Contract with the Earth is an important book at multiple levels. In simple and clear language, it conveys what all people, and Americans in particular, need to know about the environmental crisis: the nature of the problem, the cost of inaction, how to fix the problem, and the many benefits that will follow if we take action now.

Newt Gingrich and Terry Maple write as both realists and visionaries. They define the environmental challenge as far more than just averting disaster. They envision it as an opportunity to strengthen America, to sustain its economic and intellectual leadership, and, not the least, to unify it in moral purpose. They see environmental solutions as eminently achievable by the democratic process. Green is good, they say, when it serves the commonweal in accordance with the ideals that made this nation great.

It may seem odd to some that a prominent political conservative should generate such a vision. But Gingrich, accompanied by Maple, is not departing from conservatism; he is restoring it to its original, full meaning. Both "conservatism" and "conservation" are derived from the Latin *conservare* because each is meant to convey the ideal of saving and treasuring that which has, over time, proved best for humanity.

Written at a time when the United States is so widely—

and often needlessly—riven by ideological disputes, proper attention to the environment can be a unifying effort. It favors no religion or ideology; it offers both short-term and long-term benefits for all. And it can heal.

Edward O. Wilson

PREFACE

Why We Are So Committed to

A Contract with the Earth

Terry Maple and I share a lifelong passion and persistent curiosity about the natural world. In our childhoods, separated by the breadth of our great nation, in California and Pennsylvania, respectively, we spent countless hours outdoors examining the wealth of small creatures living in our midst. We explored the hills, canyons, lakes, and streams near our homes and pestered our parents to take us to the region's natural history museums, nature centers, and zoological parks. In such places, we found inspiration, knowledge, and perspective. The environmental philosophy we espouse derives from an enduring respect for wildlife in all its splendid diversity. We are personally diminished by the loss of each and every species or essential habitat that cannot resist extinction. As I travel the globe, I continue to visit local national parks, reserves and designated wilderness habitats, natural history museums, botanical gardens, aquariums, and zoos. Even when joyfully in the company of charismatic wildlife, I am reminded constantly that our failure to resolve serious environmental challenges will compromise the lives of our children and our grandchildren. However, if we engage our citizens and pull together in the same direction, I am confident we will avert a catastrophe and successfully renew the earth to its natural condition of abundance and vitality.

Environmental stewardship is everyone's responsibility, including Congress's. That is why I worked so diligently as Speaker of the House to protect the Endangered Species Act, historic legislation that has been mired in some controversy. Despite its flaws—and there are some—it is an essential conservation tool. In my evaluation of the legislation, I sought the advice of leading biologists such as E. O. Wilson and Thomas Eisner, esteemed members of the National Academy of Sciences. Good government depends on the counsel of our nation's best scientists. The Endangered Species Act is an excellent example of the value of civility, consultation, and collaboration. Mediation and compromise by interior personnel in the field (by contrast with office bureaucrats) have produced good results, a function of shared values and democratic ideals.

A Contract with the Earth presents stories, ideas, and events that illustrate how people worldwide are coping with environmental problems. It is not an exhaustive account; thousands of examples of successful environmental innovation and enterprise exist. A day surfing the Internet reveals encouraging situations and new opportunities. The entrepreneurs and grassroots and civic leaders who courageously address this challenge are the source of my continuing optimism.

An early example of private-public partnerships emerged in Atlanta during the late 1980s, with the Atlanta Zoo's transformation from a rundown failing city zoo into a privatized, world-class facility rebranded Zoo Atlanta. Atlanta's local government recognized the need to restructure the zoo as a nonprofit organization with the assistance of Atlanta's entrepreneurial business community. Zoo Atlanta quickly became one of the most successful conversions of a zoo from government to nonprofit management. Terry Maple served as the founding president and CEO, responsible for administering the reforms of the new private-public partnership. He served nearly two decades as an entrepreneurial environmentalist, battle-tested on the front lines of conservation. We both recognize that world conservation will benefit from merging private and public interests, as entrepreneurial partnerships are incorporated to serve the common good.

Americans must reach a broad-based agreement on the environment. Adversarial politics has prevented a strategic consensus from driving our nation's environmental vision. As a result, we have become a conflicted, confused, and timid polity when it comes to environmental concerns. Historically, America has been a decisive nation. We must now take the necessary steps to return our country to a position of leadership on the environment. It is not too late to make a difference. Although I was trained for a career in academe, I've spent my entire adult life immersed in politics. I am convinced, however, the environment is an issue that transcends politics. Americans deserve candor on this subject: why the environment is so important to all of us, and why the time has come to act on what we know.

Environmental leadership requires the ability to look beyond stereotypes. Environmentalists are not exclusive to one political philosophy. It is quite possible to be a green conservative; indeed, a conservative philosophy is highly compatible with the mainstream, or entrepreneurial, environmentalism that Terry and I advocate. I am not surprised that liberal thinkers have also promoted green enterprise. Business is no longer regarded as an adversary to a clean environment. Rather, global industries are the source of brilliant, workable solutions to vexing environmental problems. There should be no exclusivity when it comes to the environment; everyone is welcome at the table, and everyone is needed.

No single enterprise, event, or idea will renew the earth. Instead, I believe it will take a movement composed of dedicated citizens who can see the world in a new way and who will work together to bring about revolutionary changes in the way we conduct our lives. A Contract with the Earth aims to inspire a commitment to protect the integrity of the natural world and to usher in a promising age of environmental problem solving.

I will continue to articulate and address these issues in my public life, and I invite you to participate in a continuing dialogue. Please tell us your story share your thoughts and experiences by going to www.contractwiththeearth .com and work with us to restore and protect the good, green earth that our Creator has provided.

Newt Gingrich

ACKNOWLEDGMENTS

This book is a collaboration that evolved over many years. We gratefully acknowledge the intellectual contributions and social support of many colleagues and friends throughout the world who have helped us to assimilate knowledge and to focus our passion for the environment. We are especially indebted to Dr. Wayne Clough, who provided insight into the frontiers of environmental technology and access to cutting edge research unfolding at the Georgia Institute of Technology. Provost Gary Schuster, Vice-Provost Anderson Smith, Professor Ed Loveland, Professor Lawrence James, and Professor Bryan Norton also provided advice, guidance, and new opportunities to learn. We are also grateful to former colleagues at West Georgia College who helped make the interdisciplinary environmental studies program a distinct success and a profound influence.

We appreciate the access to many zoological parks, aquariums, wildlife facilities, national parks, and reserves where we were inspired and humbled. Resourceful gatekeepers and interpreters included Zhang Anju, Mike and Joyce Basel, Gary Clarke, John Coe, Bill Conway, Jim Else, Ron Forman, Bill Foster, Clayton Freiheit, David Hancocks, Jack Hanna, Gary Lee, Don Lindburg, Keith Lovett, John Lukas, Cynthia Moss, Gunther Nogge, Stephanie

Powers, Craig Piper, George Rabb, Ramon Rhine, Dietrich Schaaf, Christian Schmidt, David Towne, and Cunningham von Somaren, to name an important few.

Our opportunities to learn about the environment were also aided by our working relationships with Kane Baker, Nelly Bourne, Syd Butler, Paul Coverdell, Carolyn Boyd Hatcher, Robert Holder Jr., Zell Miller, Robert C. Petty, Clare Richardson, Lessie Smithgall, Joe Tanner, Mark Taylor, Jean Wineman, Andrew Young, and Zhang Zihe. Special insight into scientific issues has resulted from our conversations with Edwin Colbert, Rita Colwell, Howard Cramer, Dick Dangle, Joe Erwin, Jack Horner, Michael Novacek, Mark Norrell, Robert Sommer, Richard Tenaza, Lynne Rothschild, Bob Walker, Paul Walker, and E. O. Wilson. We also gratefully acknowledge the lessons learned from studying the published works of Roy Chapman Andrews, Henry Beston, Raymond Ditmars, Heini Hediger, Aldo Leopold, Henry Fairfield Osborne, Gifford Pinchot, and Robert M. Yerkes. Time spent at Calloway Gardens with Bo Calloway and his environmental team has been both inspirational and informative. We also acknowledge the decisive environmental leadership of Jim Jacoby and Hilburn Hillstad whose experience in innovative brownfield mitigation and landfill energy systems demonstrates what can be achieved with a total commitment to innovation, partnership, and entrepreneurial spirit.

We are indebted to our primary editor at the Johns Hopkins University Press, Vincent Burke. The comments and constructive criticism of Vince Haley, Rick Tyler, and Daniel Ballon and the editorial skills of Andre Barnett were extremely helpful. The business acumen and consistent leadership of Kathy Lubbers greatly facilitated our collaboration.

Even as we acknowledge our debt to the many friends and mentors who shaped our ideas and our point of view, we take full responsibility for any mistakes, omissions, or misinterpretations herein.

ARE YOU A MAINSTREAM ENVIRONMENTALIST?

Rate yourself on a scale from 1 to 10 for each statement below, with 10 meaning that you strongly agree with the statement and 1 meaning you strongly disagree.

_____ A healthy environment should be able to coexist with a healthy, growing economy.

_____ Investments in science and technology will generate solutions to most of our environmental problems.

_____ Incentives should be offered to encourage corporations to clean up the environment.

_____ Most disagreements about the environment can be resolved through the art of compromise.

_____ Governments can play an important role in fostering and incentivizing a healthy environment but lose support when they are too controlling.

_____ Democracies have been far better environmental stewards than totalitarian states.

_____ Corporate and private philanthropy is essential to the success of a global environmental movement.

continued

_____ Most of us have been taught to respect and protect the natural world.

_____ Political leadership will be defined in the twenty-first century by having a strong commitment to environmental stewardship.

_____ America must be a global leader on environmental issues.

If you scored higher than 70, you qualify to be labeled a "mainstream environmentalist."

A Contract with the Earth

I
Framing the Contract

What lies behind us and what lies before us are tiny matters compared to what lies within us.

RALPH WALDO EMERSON

WHETHER WE LIKE IT OR NOT, humanity has assumed responsibility for the welfare of the earth and all the noble creatures that share it. The scale of human civilization, the volume of our economic activity, and the power of science and technology have made us shapers of much of the earth. The power to shape leads inevitably to a responsibility to wield this power wisely and carefully. America, as the world's sole superpower, is obligated to provide environmental leadership at a time when so many world leaders are wringing their hands at the sheer enormity of the task.

Americans know that shaping a healthy environment is the one challenge that eclipses all others. Without a green and productive Earth, clean air to breathe, and healthy streams, rivers, lakes, and oceans, life as we know it cannot survive. But where is America's environmental playbook? And who has vetted the principles that will form the foundation of our strategy and the metrics by which success or failure can be measured? Who has the proper balance of courage and expertise to lead? *A Contract with the Earth* offers a new approach to the challenges of the twenty-first century, encouraging our citizens to accept the responsibility of global environmental leadership and to overcome our nation's troubled history of vacillation and withdrawal when the opportunity to lead beckons. We offer this preamble to *A Contract with the Earth* as a platform to frame our commitment to renew the living earth.

OUR COMMITMENT

1. **Demand Objectivity.** To achieve a fully functional and ultimately sustainable world, we must forge a cooperative working partnership among scientists, entrepreneurs, and our nation's citizens. To accomplish this, an empirical standard of objectivity should be reaffirmed. Polarized political factions have spent decades fighting over the meaning of scientific findings often distorted by the media. Despite many challenging environmental problems, there is reason for optimism, but today's science will not save us; we must strengthen our investments in science and technology to enlarge our environmental tool kit and prepare for any outcome. Good public policy requires accurate, objective science and a trustworthy and transparent process of dissemination. We must return to a civil society where candid discussion, debate, and disagreement result in rational compromise and consensus.

2. **Educate and Inspire.** We must face up to the challenge of recruiting, training, and equipping a nation of budding entrepreneurial environmentalists. The complexity of the earth's dynamic systems challenges our best and brightest scholars, so we should commit our nation's resources to recruit a generation of well-trained scientists and engineers to address environmental issues. As Soviet success in space challenged America in 1958, the urgent priority and moral cause of the environment should help to upgrade and reform our nation's underachieving pub-

lic schools. A new generation of computer-savvy, creative kids may yet be empowered and unleashed to repair, restore, and renew our resilient planet. We should hone the tools of environmental stewardship to inspire, mentor, and serve the next generation and beyond.

3. **Encourage Green Enterprise.** To protect and sustain Earth's vitality, we must endeavor to calculate, monitor, and control humanity's footprint, finding workable solutions to reverse the historical trend of waste, pollution, depletion, and degradation. A generation of doomsday scenarios has failed to activate needed reforms. Entrepreneurial spirit is a more hopeful and effective pathway to reform, as our citizens are more responsive when they are encouraged and rewarded to discover, implement, and export new technologies and better industry and government practices. Free enterprise is not the enemy of the environment; it is the engine that will drive promising alternatives to failed practices. We can think of no better place than America to establish, nurture, and expand green enterprise and innovative environmental technology.

4. **Give, Help, and Share.** It is time to export vigorously American know-how and technology through business, foreign aid, and strategic philanthropy. Investments in new technologies will contribute to vast improvements in the world's health and welfare. Wealth creates demand for cleaner, greener communities, which is especially important in former and current totalitarian states where the environment has never been a priority. Policies that pro-

mote and sustain a growing economy best enable the United States to make a national commitment to a healthier, cleaner environment and generate one of the leading industries of the twenty-first century. Economic growth and green development are highly correlated throughout the world. We have a compelling responsibility to share our ideas with other nations as America is better prepared to invent solutions for the world's environmental problems than any nation on Earth.

5. **Think Long Term.** A serious commitment to renew the earth requires that we think long term about the environment. Engendering serious change will require strategic thinking and decision making on a grand and unprecedented scale. The 2008 presidential campaign should challenge our nation's leaders to articulate a new and comprehensive vision for environmental problem solving. By applying the environmental mantra "think globally, act locally," we can engineer solutions to environmental problems tailored to fit local circumstances and avoid heavy-handed, federal programs that threaten the autonomy of towns, cities, and states. To secure the future of the earth, we must train ourselves to think long term and look for creative solutions "outside the box."

ADVANCING WITH CONFIDENCE AND OPTIMISM

These five principles guide our Contract with the Earth. Set forth in the next chapter, the contract consists of commitments that help to define an evolving strategy for protecting and enhancing the natural world long into the future. Adhering to these principles and commitments, we anticipate a nation more comfortable with the obligations of world leadership. The nation's confidence is uplifted by our advanced technology, a tested work ethic, optimism, and an unmatched entrepreneurial "can-do" spirit.

We understand that our commitment to renewing the natural world requires value judgments based on some standard of quality. We cannot discuss these issues without judgments of this kind, but we also know that standards of quality ultimately rest on objective and accurate quantitative measurements that are independent of values. We agree that science gathers data in a rational, systematic, testable, and reproducible manner, as biologist Robert Lackey has recently asserted, and we affirm his contention that scientists should be sensitive to the boundaries between science and value judgments. Lackey's warning to policy makers is worth repeating:

Call our hand when you observe us overstepping our role as scientists and slipping into stealth policy advocacy. Scientific information is too important to the resolution of vital, divisive, and controversial ecological issues to allow some scientists to marginalize science through its misuse.

For our part, we readily acknowledge that values have played an important role in formulating this Contract with the Earth. We take our stand without hesitation. Quantitative measurements of biodiversity, ozone, lead, mercury, forest cover, fish populations, and polar ice all stand as objective indications of environmental quality. Where to draw the line on quality is the challenge that policy makers face daily. We depend on accurate scientific data to guide us, but it is "we the people" who will ultimately decide how best to achieve a sufficient and sustainable quality of life.

We recognize that a contract is both a binding agreement and a serious long-term commitment. To this end, we envision a nation of mainstream environmentalists whose commitment to safeguard the earth is unassailable. It is time for a bold initiative on behalf of the natural world, dedicated to a common cause and a bridge to green prosperity; it is time to establish and embrace a Contract with the Earth.

2
A Contract with the Earth

The sea, the great unifier, is man's only hope.
Now as never before, the old phrase has a literal
meaning: We are all in the same boat.

JACQUES YVES COUSTEAU

OUR NATION'S MORAL OBLIGATION to provide effective environmental leadership will require the formation of new strategic partnerships among nations, nongovernmental organizations, and multinational corporations dedicated to protecting and renewing the earth's precious resources. International partnerships can only be achieved with presidential leadership and significant bipartisan support in Congress. For too long, the adversarial political climate in America inhibited cooperation, consensus, and action in this domain. Civility, compromise, and unity of purpose are necessary precursors to meaningful global cooperation. In addition, the focused energy, creativity, enthusiasm, and endorsement of the American people will be necessary to achieve the gold standard of a sustainable, renewable world. To reach new benchmarks in protecting and renewing the environment and guarantee a bright and enduring future for our children and grandchildren, we ask you to join with us to advance the ten commitments of this Contract with the Earth.

1. **Take the Lead.** Let us affirm that America is willing to resume its role as the environmental leader of the world. Further, the burden of leadership requires that Americans help shape an earth where waste is minimized by reducing, reusing, and recycling; where water and air meet stringent standards of cleanliness; where fossil fuels have been largely modified for carbon recycling or replaced by carbon neutral alternatives; where forests, wetlands, lakes, and the world's oceans have been restored to health;

where the pace of plant and animal extinction has been effectively abated and biological diversity regenerated; where humanity's collective impact on the earth has been moderated by prudent management of our natural resources and innovative technologies that act to replenish and restore the earth's damaged ecosystems; and where the environmental quality of life is no longer in decline but continues to improve for all nations throughout the world.

2. **Reward a New Generation of Environmental Entrepreneurs.** Acknowledge that we must create the context in which entrepreneurial environmentalists will flourish. Reject the notion that free enterprise and environmentalism are opposing forces. Stoke the competitive fire of environmental science and green enterprise through significant investments that hasten the pace of change and innovation. Focus and nurture environmental business, large and small, throughout the nation, to create and advance a multitude of renewable, sustainable, and restorative technologies. Continue the momentum that has made America a world leader in energy efficiency by providing powerful incentives to cultivate genius and innovation. Reach out to potential collaborators to form diverse and inclusive partnerships to disseminate prosperity and opportunity. Significantly expand the pool of environmental entrepreneurs by ensuring that higher education is available to every American who wants to learn.

3. **Retire or Rejuvenate Old Technologies.** Diminish our dependence on environmentally unfriendly and harmful

technologies by reforming government and industry, offering compelling alternatives and incentives to invest in and to transfer to better technologies. Innovation, improved industrial operating standards, and the constant migration to emerging best practices will modify old technologies into acceptable forms. Let us commit to change—and lead the change—but not retreat when faced with formidable challenges. By overcoming difficult problems, we are challenged to pay attention, think, and create for a better future. We must commit to a future characterized by ubiquitous, clean technologies.

4. **Transform the Role of Government.** Government alone cannot solve complex environmental problems, but government in partnership with organizations and businesses ensures that thoughtful environmental action will strengthen both the economy and the environment. Government, at all levels, should be a facilitator for entrepreneurial, private-sector innovations and the formation of private-public environmental partnerships, supporting and not suppressing the creativity of entrepreneurial environmentalists. The resources of government should be consistently applied to reduce red tape and facilitate progress. Some regulation will always be necessary, but it should be limited, focused, and reasonable to liberate the full potential of market innovation.

5. **Become an Aspirational and Inspirational Nation.** Expect that our local, state, and federal governments and their business enterprise partners will be both aspirational

and inspirational. Government, industry, and small businesses should advance lofty but achievable goals to improve the environment step by step. The arbitrary power of our federal bureaucracy should not impose or dictate environmental aspirations. Instead, they should emerge from shared values and a consensus established by strategic dialogue among a broad array of stakeholders. We have everything to gain by thinking big and aiming high. Our aspirations will drive us; an inspired political, intellectual, and social process will guide us. To inspire others to achieve, we must continue to express optimism.

6. **Position America to Meet the Challenge.** Require our leaders to formulate positive, pragmatic, and proactive public policies so that looming environmental crises, whether near term or distant, can be identified and averted. We should strive to achieve an enduring consensus on the specific goals for a healthy environment and agree on the policies to achieve them. Leaders should reach out to every citizen and to more than one-half-million elected American officials at all levels of government to chart a course together. We must be prepared to anticipate and quickly respond to present and future threats. The high priority of the environment must be affirmed.

7. **Encourage Scientific and Technical Literacy.** A dramatically larger workforce of environmental engineers, scientists, and technically competent entrepreneurs is needed to generate new ideas and new solutions in a world

of increasing complexity. Our community schools must return to an emphasis on math and science to educate citizens with the ability to fill this need. New learning technologies and a commitment to widespread and consistent mentoring will help us reach this goal. We must provide the financial incentives to encourage the development of young environmental scientists and engineers. The National Science Foundation and other relevant federal agencies should be encouraged to "think big" about reforms in public learning. Our nation's museums, gardens, zoos, and aquariums should be empowered to help stimulate and inspire young people to choose careers in science, math, and engineering and to keep them focused on the priority of the environment.

8. **Invoke the Spirit of Collaboration and Cooperation.** A history of adversarial politics and litigation has deferred, delayed, and deterred an adequate response to our growing environmental challenges. Antienvironmental politicians are out of step with the American people as concern for the environment is widely acknowledged as an important component of a patriotic worldview. A mainstream, nonpartisan approach to environmental problem solving will engage America's citizens in active, pragmatic change without political polarization. We should question how, not whether, the earth can be renewed. Politicians should compete on the basis of who has the best solutions and be judged by the outcomes of solutions they sponsor.

We should agree to elevate the environmental debate to a higher plane of civil dialogue. A cessation of shouting is long overdue; it is time to communicate.

9. **Support the Environment through Philanthropy and Investment.** A coordinated, strategic philanthropy will support the increasing priority of environmental events and issues. We need to enlist America's most affluent corporations, foundations, and individuals to help solve complex environmental problems. The historic generosity of Americans must be encouraged as it continues to be a major source of our strength and reputation throughout the world. Strategic philanthropy will be an essential tool of entrepreneurial environmentalism in the twenty-first century.

10. **Enlist the Nation.** It is time to recruit an army of environmental foot soldiers to tenaciously pursue a new course for our nation. In addition, executives in government, business, science, and the arts must rally to mobilize all citizens to pursue proactive, environmental policies and practices at home and in the workplace. If America dares to lead on the environment, our elected representatives—at local, state, and federal levels—also must be fully committed to the task. This commitment of time and energy is nothing less than a quest to restore trust, teamwork, and cohesion to our nation as we engage a new and comprehensive environmental agenda. Every one of us, meek and mighty, is needed to reach our goal of a cleaner, healthier Earth.

IMPLEMENTING A CONTRACT WITH THE EARTH

In the chapters to follow, we will amplify and explore the ten commitments of the contract. Because promoting a vital environment will always be a work in progress, Americans will need to stay engaged in the same way that we focus daily on the health and welfare of our families and the safety of our communities. We hope the contract will facilitate daily conversations about the environment and stimulate new ideas and new information that will lead to new solutions. The examples we present are illustrative but not exhaustive because, literally, thousands of stories are worth telling.

In the months and years ahead, we intend to stay in touch with motivated, entrepreneurial environmentalists like you. After you have read and thought about the contents of this book, we ask you to contact us at www.con tractwiththeearth.com so you can share your ideas, offer constructive criticism, and discuss the environmental news of the moment. You are the key to solving the earth's environmental problems, and your community is the best place to affirm your commitment to a better world. We look forward to an ever-expanding dialogue with the American people and all others who share our approach to environmental problem solving.

The challenge ahead is serious, more a marathon than a sprint, but it may be helpful to think beyond ourselves to commit to membership on a global team that is thoroughly green—a team that transcends political party affiliations or ideological agendas. Cohesion, cooperation, and collaboration are critical features of the new century's environmental playbook. If we succeed in mobilizing our nation, our experience may become a model for other nations. We may also succeed in uniting the world in a shared mission to shape a green, clean, and safe planet with liberty, justice, and prosperity for all.

3
A Matter of Respect

Respect. (n) A feeling of high regard, honor,

or esteem.

WHO AMONG US LACKS A FUNDAMENTAL respect for the earth? Without respect, we would not bother to recycle or care for our gardens or brake for the errant squirrel. Parents universally teach their children to respect the environment, encouraging them not to litter or waste energy. We want our schools to reveal the many interconnections and interdependencies between the environment and our personal experience. For most of us, respect for the environment is not some far-out concept; it is an embedded value of mainstream America. Among the college students we have taught in Georgia colleges and universities, we were gratified to find near-universal expressions of respect for the natural world. From kindergarten to college, it is easy to teach respect and concern for the environment.

In childhood, family visits to museums, zoos, aquariums, botanical gardens, national parks, monuments, and wilderness reserves inspired us. The palms, peccaries, porpoises, porcupines, pythons, and other oddities we encountered in these natural settings sparked our youthful imaginations. These venerable institutions and venues still harbor a huge potential to inspire and educate young people; yet, they have been underused and underappreciated as community centers for inspiring learning, shaping attitudes, and changing behavior. Because they so effectively inspire and teach, our museums, gardens, and zoos should receive our full support.

Family trips to a museum or a zoo are always enlightening and a lot of fun, but we are fortunate to be surrounded by unbridled nature in our own backyards. If we just bother to look, we can see a vast population of fascinating local plants and animals in and around our homes. Our neighborhoods may be regarded as "community preserves" that most of us help to manage wisely if not diligently. Helpful environmental groups such as the National Wildlife Federation provide useful information on how to assemble and manage our backyard habitats. We eagerly create them at home, and we build them on the playgrounds of our neighborhood schools. The nearness of nature is an unavoidable and wonderful fact of life. It is here that respect is nurtured on a daily basis.

Some of the fauna in our yards instantly evoke respect. Butterflies, for example, have a special capacity to enchant and inform. Callaway Gardens, a botanical oasis just outside Atlanta, is visited annually by hundreds of thousands of people who want to experience the charismatic nature of these frail, enigmatic creatures at the Cecil B. Day Butterfly Center. The blinking blanket of color in constant, silent movement eventually captures the imagination of even the grumpy few dragged along by more adventurous members of the family. Children can never get enough of the natural world. Butterflies have other virtues; scientists have discovered that butterflies are highly sensitive to changes in the environment. A healthy population of butterflies is a sign that all is well; a precipitous decline is a sign of danger ahead. Respect and awareness are easily cultivated within a butterfly center or garden. An early exposure to butterflies will surely influence future botanists and zoologists, but the collective impact of these charismatic insects on humanity at-large is a greater and more enduring legacy. We know firsthand the potency of charismatic wildlife.

Americans born after World War II grew up in a land and time that encouraged a heritage of optimism about the future. Doomsday scenarios depicting looming environmental crises and disasters are starkly out of sync with such a state of mind. We argue that they should be. The important lesson of the past few decades of environmental awareness is that the interests of wildlife and the environment are better served by optimism and hope. The environmental challenges are real, but our imagination and innate creativity give us confidence that humanity, against all odds, can and will prevail. The overwhelming complexity, elegance, and grandeur of nature powerfully motivate our concern and our resolve to protect these priceless assets.

The esteemed ecologist Garrett Hardin, in his benchmark 1968 essay, "The Tragedy of the Commons," warned that the earth and its life-forms were in grave peril. He argued that unrestricted access to natural resources and short-term thinking leads inevitably to depletion of the earth's limited resources. Consumers of seafood know the personal price when the tragedy of the commons is played out in economic terms. When shrimp, redfish, crab, and abalone are overexploited, their price increases, and the commodity is no longer available in the marketplace. The near loss of the Atlantic cod supply in the twentieth century is a painful reminder of the consequences when we fail to monitor and protect our resources. In some settings, privatization, an alternative to the concept of "commons," might lead to more accountability by the owner of a limited resource. As Hardin argued forty years ago,

> Maritime nations still respond automatically to the shibboleth of the "freedom of the seas." Professing to believe in the "inexhaustible resources of the oceans," they bring species after species of fish and whales closer to extinction.

In Hardin's original essay, the oceans form a "commons"—a place open to use by all. The ocean, unlike land, is not a place where a family or company has a vested interest in future productivity. Hardin suggested (and has yet to be proved wrong on this particular point) that the advantages to the individual of overexploiting a commons are much less than the disadvantages to the individual. Yet, the disadvantages to society (eventual loss or diminution of the resource) are much greater than the advantages (a greater supply for a brief period).

Hardin's solution to the problems of environmental overexploitation, however, does not ring true. He suggested the option of "relinquishing the freedom to breed." In an otherwise philosophical essay, his solution strikes us as disturbing, impractical, and highly undesirable. Families without children, with a couple of children, or with any number of children can all be good stewards of the environment—and our premise applies to all families in every nation. Indeed, in some respects, the population problem is solving itself, with birth rates falling as nations develop healthy economies with stable, predictable futures.

Overpopulation is a problem we can handle most effectively by targeting foreign aid and encouragement for emerging democracies with a stable rule of law and growing economies. Poverty and population explosions are highly correlated. We have learned much in the past four decades about the relationship between increased wealth and decreased family size and about the many ways we can reduce our environmental impact. As the "green" economist John Baden has argued, as people become better educated and wealthier, they demand an environment of superior quality. "In general," Baden concluded, "richer is greener." This is the principle that links the economy with the environment.

Effective economic policy is both a profound humanitarian act and a major step toward a better environment. By helping to guide developing nations into a more prosperous economic future, we will circumvent sixty years of failed economic policies in the world's poorest countries, and we will likely avoid further environmental degradation in the process. Thirty years of research, stimulated by

Hardin's thinking, reveals that tragedies of the commons are real but certainly not inevitable. As ecologist Elinor Ostrom and her colleagues asserted in a 1999 article in *Science,*

> building from the lessons of past successes will require forms of communication, information, and trust that are broad and deep beyond precedent, but not beyond possibility.

Harvard biologist E. O. Wilson, who described his concept of *biophilia* two decades ago, offered another optimistic perspective on our relationship with the environment. Wilson theorized that there is an innate bond between humanity and nature. According to Wilson, people have a deep, even subconscious, affiliation with the natural world. It is a basic feature of our human biology. Most of us can sense this subtle feeling: during a walk along a beach, at rest in the glow of a spectacular sunset, or in the presence of majestic wildlife. Not surprising, Wilson, trained in entomology, experiences that feeling when he is watching ants. The phenomenon of biophilia explains why it is relatively easy to evoke sympathy for the environment and why we are filled with guilt when we damage or denigrate it. We seem to be endowed by our Creator with interest in and a sense of obligation to care for the natural world.

If caring is a natural propensity of humankind, caring

can also be induced by contact with powerful and compelling naturalistic stimuli. None but the most jaded among us can avoid the feeling of awe that is engendered by visions of the Serengeti Plains, the oases of Okavango, the vast seclusion of the Okefenokee Swamp, the massive and majestic North American glaciers, or the volcanic forest habitat that is home to the world's last remaining population of mountain gorillas.

Is there an American—whether right, left, or center—who denies that such places must be protected? We know that our fellow citizens have profound respect for these and the thousands of other iconic natural places on Earth, even if they haven't visited such places. Yet, as generous as we can be when earthquakes, hurricanes, and tsunamis strike in remote locations of the world, our response to the myriad threats to endangered animals and plants or the glorious habitats on which they depend is too often tepid. Do we exert the resolute leadership that demonstrates our respect for the natural world and the species that live in it? Regrettably, our behavior does not always reflect the strength of our beliefs.

Our inertia belies the fact that we fit the description of "mainstream environmentalists"—we don't wear any special political armor; rather, we share the core values of appreciation, respect, and active stewardship for the earth. This grand coalition of parents, neighbors, pastors, teachers, and students, virtually every kind of person, is surely the largest interest group in America, but we must be activated by events or by inspiration. In this way, we are a kind of "silent majority." Family and future oriented, the sage advice of Theodore Roosevelt still resonates with all of us:

> The nation behaves well if it treats the natural resources as assets, which it must turn over to the next generation increased, and not impaired, in value.

Six decades after Roosevelt correctly connected American prosperity with American environmentalism, we have witnessed a host of new problems and opportunities. Somewhere in our zeal to produce and achieve, we lost sight of Roosevelt's insight and perspective. What would this great leader confide at the dawn of this environmental century? Surely, he would counsel commitment, cohesion, and courage. In the tradition of his vigorous leadership, we must face up to our responsibilities, roll up our sleeves, and get to work. With time and effort, the silent mainstream may become a vocal majority and a force to be reckoned with in the environmental arena.

We are a wealthy nation, and some have argued that our wealth has led us to a state of arrogance and disrespect for the earth. This argument does not ring true. Our high-powered economy is sufficiently sophisticated and flexible to permit coexistence with a healthy natural world. We willingly create parks, monuments, and refuges; pass laws against pollution and the extermination of species; and live lives that demonstrate our appreciation for the value of ecosystems. Americans have expressed their respect for the earth in poll after poll. We have actively chosen to both live in the modern world and to avoid the tragedy of the commons. But if we aspire to a position of global environmental leadership, we will need to quicken our response to environmental emergencies, consistent with the goal of delivering to our heirs a world "increased, and not impaired, in value." Our business acumen and entrepreneurial spirit must be used in service of achieving a "more perfect" stewardship of the earth.

If we approach the environment with the zeal and spirit of an entrepreneur, we can accomplish much more than just the application of some bureaucratic band-aid. A promising example of the new environmental approach is a unique training program in environmental entrepreneurship established at the Foundation for Research on Economics and the Environment in Bozeman, Montana. The program is designed to "explore the creation of new institutions and innovative public policies that promote environmental progress." Commentary in the newsletter FREE is refreshingly relevant. Entrepreneurial approaches to the environment are being honed in the West. It will be a major breakthrough when such programs are adopted by traditional business schools.

Although the label was not so widely used in 1970, we can vividly recall scores of entrepreneurs who invigorated the topography of the first Earth Day. Vendors hawked various new Earth products and publications to a curious population of largely college students. The prescient *Whole Earth Catalog*, published in 1968, was a gold mine of useful information that also emphasized alternative forms of energy. Entrepreneurs are optimists by nature, and they are not easily intimidated. In government and in industry, entrepreneurs who bet on big ideas such as renewable energy will eventually prevail over prophets of doom that buffer the status quo. We expect an Earth Day metamorphosis in the next few years as more people enthusiastically sign up for a revitalized, entrepreneurial environmental movement. Economists ought to revisit the symbolic power and imagery of Earth Day.

Corporate America is also evolving as it adjusts business practices to embrace conservation. Many American banks, for example, have committed to the World Wildlife Fund's Equator Principles, limiting investment only to companies and projects committed to environmental protection. Wal-Mart now only sells fish it has bought from sources that practice sustainable harvesting methods, as certified by the Marine Stewardship Council. Increasingly, business leaders regard our forests and wetlands as ecological assets that must be protected to support human life, public positions that their mainstream customers applaud.

An excellent example of America's new environmental enterprise is demonstrated by the experience of media mogul Ted Turner who invested some of his vast fortune to ranch the American bison, an ungulate adapted to the dry western ecosystem in Montana. Bison are easier on the land than cattle and are a source of leaner, healthier meat. Turner owns about 40,000 head, more bison than any person or government. He buys western land and devotes a portion of this property to his growing population of bison. Recently, he opened Ted's Montana Grill, a restaurant chain with locations in many states that serves primarily bison burgers, a delectable sandwich, stimulating the market for his innovative meat products.

Bison has not replaced beef by a long shot, but Turner's new enterprise is a creative response to the challenge of gently using the natural assets of the western frontier. No one forced Turner to ranch bison instead of cattle, but by producing a healthy, ecofriendly dinner staple, he provided a uniquely sustainable solution to the problem of farming arid land. Similarly, African entrepreneurs have ranched local forms of livestock, Ankole cattle, for example, associated with the Tutsi tribe in Central Africa, and other local wildlife to protect their soil and plants against the harsher effect of imported beef cattle. Throughout the world, the preference for indigenous and suitably adapted local livestock has proved to be a better ranching practice as these animals can use sparse vegetation and require less water to survive.

In 1980, the membership of the Association for Zoos and Aquariums (AZA) affirmed that conservation would become their foremost priority. The logic that underlies this consensus is simple. First, zoo professionals are motivated by a deep emotional bond with the animals in their care. They are committed to the welfare and survival of all forms of wildlife. Our nation's zoos could not in good conscience exhibit wildlife without a serious commitment to saving the habitats that sustain them in nature. To promote this cause, the members of AZA created a new vision for their members and patrons, symbolized by a memorable slogan crafted by scholar-marketers at the Minnesota Zoo, arguably the simplest and most eloquent conservation slogan articulated by any zoo:

> Strengthening the bond between
> people and the living earth.

Given that 143 million people visit accredited zoos and aquariums each year, the Association of Zoos and Aquariums is entitled to the claim that the association is building North America's largest wildlife conservation movement. Zoos and aquariums are mainstream environmental organizations. Their supporters, some 48 million registered members, are extraordinarily committed to conservation, and AZA zoos and aquariums back up their boast with money, spending some $250 million in 2006 on 1,719 conservation and science projects in ninety-seven countries and regions.

Strengthening the bond of people with the earth is the cornerstone of conservation. Although our affinity toward the earth is probably hardwired, as E. O. Wilson has asserted, that affinity must be reinforced in our school systems and promoted in our national and local media. We must work at the task to be effective conservationists.

How does the environmental mainstream prepare to enter into a Contract with the Earth? The principles and commitments of the contract reflect values that most Americans share but which have not been fully articulated elsewhere. The contract has been designed to benefit all living things in a spirit of inclusive citizen dialogue and debate; it is a call to opportunity, action, and implementation. It is not a sermon or a lesson but, rather, a promise to act on a set of shared beliefs. It marks a moment where America accepts the responsibility of global leadership on the environment, consistent with our national character and our history of entrepreneurial activism. It emphasizes that American leadership will champion advanced technology, scholarship, the free market, and a renewed enthusiasm for environmental research and development. If America is true to its national character, our great nation will seize the opportunity to lead the world into a better future.

The Contract with the Earth is a tool to confront the forces and events that threaten the environment. Such threats should not be denied; they are real. Americans are prepared by their history to respond to threats to the environment. We are a nation of individuals who willingly resort to teamwork when we face adversity. Historically, our forefathers consistently met and defeated formidable obstacles to a young, democratic nation. The wagon train and barn raisings are examples of team-building experiences integral to America's history. Early in the twentieth century, our nation bounced back from a frenzy of deforestation, and we instituted reforms to save farming in the American heartland. We know how to work together to achieve a better future; however, to lead the world, given a universe of demanding environmental priorities, we will have to first put our own house in order.

Although our nation's international leadership has been expressed in countless ways, when it comes to the environment our leaders have been timid and restrained. Other nations and groups of nations have attempted to lead and beckoned America to follow. They offered the Kyoto Accords that were rejected by resolution in the U.S. Senate on a vote of 95-0. The Kyoto document is clearly flawed, and Kyoto remains unratified by the U.S. government. Something should be done to achieve sensible reductions in greenhouse gas emissions, but our government has not yet demonstrated the necessary leadership to create a workable alternative to Kyoto. Our country needs to get back to the table. If a better international approach is negotiated, all nations, including developing economic powerhouses such as China and India, must be included. Leadership, therefore, is the highest priority in our Contract with the Earth.

Make no mistake, the entrepreneurial model we propose will encounter determined opposition and bureaucratic inertia. Opposition aside, the logic of this Contract with the Earth will win. It has to win because the current environmental logjam and dearth of American environmental leadership is not an option. We lead because we must. It is our calling as Americans and our duty as patriotic, mainstream environmentalists.

TALKING POINTS

1. Respect for the natural world is a mainstream value of all Americans, but the attitude must be nurtured in our children and affirmed throughout our society.
2. Garrett Hardin's "Tragedy of the Commons" suggests that ownership, or perception of ownership, facilitates conservation.
3. We acknowledge the relevance of the concept of biophilia and the innate bond that exists between humanity and nature.
4. It is tragic that the American government, both Congress and the president, has thus far failed to exert sufficient and effective leadership on the environment. We anticipate a return to assertive American leadership.

4
Missed Opportunities

Life is short, the art long, opportunity fleeting,
experience treacherous, judgment difficult.

HIPPOCRATES

INFORMATION TECHNOLOGY HAS profoundly and permanently altered the world that molds and shapes our behavior at home and in the workplace. Today, we communicate with lightning speed, experience the news 24/7, and live in a house with jumbo-sized talking heads. It's a totally different world compared to the subdued community of our youth; there is a lot to like about it, but sometimes we loathe it. After all, office technology was supposed to reduce our daily workload, but it has clearly created more work. Our portable and digital office, for better or worse, now travels with us wherever we go. Information is ubiquitous, but our advanced information technology is the reason we know so much about global climate change and the decline of species, wetlands, and forests throughout the world.

Technology permits society to monitor closely an entire universe of physical, chemical, biological, and meteorological variables, including the world's capacious network of oceans. In California, for example, scientists have measured air temperature, wind speed, nutrient levels, salinity, water temperature, and the quantity of larval fish and zooplankton on the surface and deep below the surface of coastal waters. Incredibly, as water temperatures have risen over the years, the biological foundation of the Pacific food chain has crashed. Like technology, ecosystems are linked, and the creatures along the food chain are also experiencing a sudden and dramatic decline. In the prestigious journal *Science*, Scripps oceanographer John Mc-

Gowan reported a drop in zooplankton in the California current of 70 percent, the largest change ever measured in plankton populations. In the thirty-two years since a 1967 aerial survey of kelp forests, the California coastline had experienced a 75 percent decline. The animals that depend on a healthy food chain suffer accordingly; the sooty shearwater, a once common predatory seabird, has declined by 90 percent, just one of many bird species in trouble. Because we are so adept at monitoring changes in the earth, information overload can quickly lead to a sense of helplessness. Although no one knows the full implications of these alarming findings, it is clearly time to rally, not rant.

We don't have a perfect understanding of Earth's systems and processes; it may be unachievable in the face of such complexity. However, we continue to develop imperfect but useful mathematical models, and we have been able to isolate physical variables in the laboratory. We should celebrate our capacity to learn quickly about such things; and we need to continue funding the search for this vital information. We may need to commit to an International Environmental Year project similar to the International Geophysical Year of the late 1950s. In that important international research venture, new measurements fundamentally changed our understanding of how the earth functions. The widespread acceptance of plate tectonics as an explanation for the earth's behavior grew directly out of this experience. Today, we need a boldly conceived comprehensive system of measurement to understand thor-

oughly our global environment. We need the tools to diag-
nose the condition of the earth, and we need to act fast
with whatever solutions we can design.

Unfortunately, we have experienced such political polarization on environmental issues that scientists are not
completely trusted. By nature, scientists are cautious about
their findings, but newspaper headlines often scream the
worst-case scenario. Scary headlines don't have to be pur-
posefully driven to have a powerful effect on the reader.
And the electronic media turns up the frequency and the
volume while it loses much of the detail behind every environmental story. Media accounts of scientific findings
require the greatest care and objectivity, but the reporters
who must write these stories are not always equal to the
task. Neither the reporter nor the reader/viewer of the
information seems to understand the nuances, the meth-
odological limitations, or the caveats that accompany a
truly complex set of findings. Advancing scientific literacy
is a worthy goal of any mainstream environmental movement or a nation that aspires to lead the world.

When scientists at the University of Oxford's Biodiversity Research Group examined the media's response to a
study of climate change, published in the journal *Nature*, they found that the news media in the United King-
dom tended to misrepresent the findings, making the
consequences seem more catastrophic and the timescale
shorter. Surprisingly, some Internet accounts were more
accurate and some were critical of the underlying sci-

ence. The writers asserted that such "polarized representations of environmental science are indicative of a 'struggle for legitimacy' between environmentalist and anti-environmentalist groups." The public's trust in science is put at risk by such conflict. When research findings are overstated or misrepresented, a reputation for accuracy can be seriously compromised. According to social scientist Richard Ladle and his colleagues,

> there is a danger that environmental scientists . . . will be seen to be crying wolf and may accordingly suffer a loss of credibility and legitimacy. Eventually this could lead to public apathy and even "conservation fatigue."

Activist science is nothing new. After World War II, a group of atomic scientists advocated for a ban on nuclear weapons. They stepped beyond their role as scientists to enter a public and highly emotional debate. Given the awesome power of the atomic bomb and its aftereffects, the nuclear scientists feared even greater damage in an encounter in which two or more warring nations were armed with nuclear technology. However, scientists can be moved by conscience to reach conclusions beyond their data, and the public suffers more confusion than commitment. The credibility of scientists requires a commitment to objectivity even when they feel strongly about an issue. In a free society, this is not a simple matter, so it is not surprising that scientists frequently offer their opinions and advocacy to the media.

At a recent meeting of the Society of Conservation Biology in 2005, the role of advocacy in conservation biology was openly debated. Some argued that scientists can only harm their credibility by open activism. A government policy maker added: "I don't mind policy recommendations from scientists, but I take them with a grain of salt." In a survey of 300 people attending their conference, 70 percent agreed that the society's journal should be encouraged to advocate certain conservation policies. This is a common view of many scientific societies. In response to the politics of environmental advocacy a call for *sound science* has been issued, a counterpoint to doomsday conclusions that have framed the debate on many fronts. Sound science should not be a euphemism for censored science, but political bias from left or right must be avoided for objectivity to prevail.

In Robert Lichter and Stanley Rothman's important book, *Environmental Cancer—A Political Disease?*, published in 1999, they conducted a scientific survey of cancer researchers and environmental activists to determine whether there were differences in the two groups' views on the phenomenon of "environmental cancer." Not only did the two social scientists find differences in media interpretations of the data, they also discovered that press reports most frequently cited the views of environmental activists as if they represented the views of the scientists. Lichter and Rothman questioned the objectivity of both the environmentalists and the media over the two decades that they surveyed the issue. Further, and more serious in their opinion, this systematic bias and resultant media hyperbole seriously distorted public policy and priorities. If this can happen in a sophisticated medical field such as cancer research, then the danger exists that discussions about other environmental issues will be similarly tainted by media bias. The distribution of carcinogens in the environment is a serious problem, but we must be able to calculate the true risks by careful interpretation of the scientific findings. At the same time, we strongly agree with the conclusion that a healthier environment is needed to defeat many forms of cancer.

Action regarding the environment requires objectivity, precision, accuracy, validity, replication, constructive criticism, and consensus. Scientists, engineers, and economists have to stay focused on putting accurate data into the hands of decision makers, while they explain their findings to the public, which, in the end, wields decisive power in a free society. Mechanisms must be developed to transform highly technical findings into governmental and economic policies.

If we can reach a point where two (or more) political parties are sincerely committed to peremptory environmental action and differ only in the details and designs, we have arrived at the brink of consensus. Currently, liberal politicians operate as if they own the issue; in their reaction, conservatives appear to disdain it. As the media overreacts to information and generates sensational headlines, mainstream America tunes out. This is the worst kind of outcome. In such a situation, if action is needed, we would be driven from action instead. Given such uncertainty, we are prone to a malady psychologists have labeled *learned helplessness*. We cope by doing nothing to change the situation. As we cope, we momentarily escape stress and anxiety, but all the while, the problem only deepens.

We must find sufficient common ground to address a universe of complex environmental problems. A prime example is a national fishing industry rife with conflict. In the West, some commentators believe the salmon industry is essentially finished. Salmon production is limited by the organism's need to use natural rivers and streams. A long history of dams, aqueducts, and other man-made barriers has restricted the free movement of fish. Dikes and levees create simplified channels rather than complex habitat, and land development leads to a hardening of watersheds, an unfriendly environment for spawning. If we remove the dams, we may expose our communities to more frequent flooding or lose access to water for energy, irrigation, and recreation. It is a difficult problem and will not be easily or cheaply corrected. An issue this challenging must be addressed with a commitment to unequivocal objectivity.

At the popular website Salmon Friendly Seattle, the local government in Washington state distributes an "Urban

Blueprint for Habitat Protection and Restoration." This document reveals how a community can protect and restore beaches and marshes as well as restore tidal flats, wetlands, and channels to support fish habitat. Once these complex issues are understood, a nonpartisan group of experts and decision makers can address them collectively and recommend a suitable action plan. All stakeholders must participate rationally. More than all other issues, the health of the earth cannot be understood or advanced by a political climate of "we versus thee." A nonpartisan approach means that no political philosophy dominates and everyone has a stake in the outcome.

In fact, Puget Sound salmon recovery efforts in Washington state have been inclusive and effective. A "Five-Step Shared Strategy" was developed because salmon recovery efforts in the Pacific Northwest appeared to be operating in isolation. The plan brought together National Oceanic and Atmospheric Administration fisheries, the U.S. Fish and Wildlife Service, Washington Gov. Christine Gregoire, Puget Sound tribes, state natural resources agencies, local governments, and key nongovernmental organizations. The coalition took the strong position that the health of salmon indicates ecosystem health and the health and welfare of local people. The foundation of the shared strategy is the belief that the "people of Puget Sound have the creativity, knowledge, and motivation to find lasting solutions to complex ecological, economic, and cultural challenges." This example illustrates the efficacy of "acting locally."

Looking back on the thirty-six years since the first Earth Day, environmental groups, government, and industry clearly failed to build on the momentum of a fledgling, grassroots environmental movement. Although the Endangered Species Act and the Convention on International Trade of Endangered Species treaty were enacted in 1973, the unity that produced these landmark achievements failed to spark substantive and sustainable change. Recycling took hold quickly, and the green product movement is becoming a substantial business, but wider achievements have not measured up to the full potential of the epoch. Regrettably, baby boomers must also acknowledge that the environment was not a priority in the administrations of two baby boomer presidents, Bill Clinton and George W. Bush. There were platitudes and a few praiseworthy achievements, but neither president succeeded in significantly advancing environmental policy.

Recently, President Bush declared the vast ocean and archipelago known as the Northwestern Hawaiian Islands as a national monument. At 1,400 miles long and 100 miles wide, the site is 100 times larger than Yosemite National Park, larger than forty-six of the fifty states, and home to more than 7,000 marine species. American presidents have the authority to make far-reaching environmental commitments. President Richard M. Nixon's environmental legacy was the Endangered Species Act. It appears that the Hawaiian Island Monument may be the environmental high point of the Bush administration, but we should expect better and much more from our leaders.

As a nation, we have yet to coalesce around a compelling environmental vision, but one promising candidate has been vetted again and again. By weaning industrial societies from their dependence on fossil fuels, the world would be a far better place. Even President Bush, a former oil man, has characterized America's thirst for oil as an "addiction." The metaphor may be appropriate since addictions are hard to break. Although the United States makes up only 5 percent of the world's population, we consume 28 percent of the world's energy. At current usage, 78 percent of our energy consumption is from fossil fuels. As the supply and price of fuels have waxed and waned (mostly waxed), the urgency to change direction has never evoked a workable solution. We know that petroleum is a limited resource, and we recognize that it is controlled by many nations unfriendly to western democracies. To sustain the world's largest economy, we need oil, but we do not control access to it. Ethanol fits the profile of a promising alternative and it represents an economic opportunity for corn producers. We are also developing cellulosic sources such as crop waste, wood, orange peelings, and grass. Soy diesel recently emerged as another successful product, while algae-based compounds have gained traction as a new frontier for ethanol investors. Each holds promise as a domestic alternative to oil acquired from totalitarian regimes.

Clearly, the big idea of alternative, earth-friendly sources of energy, such as ethanol or biodiesel, is a worthy and timely cause, but it has eluded our grasp for more than thirty years. It is easy to blame this outcome on faulty technology, but it is more likely a shortcoming of our will to act. Neither industry nor government seems to be able to launch a compelling campaign for energy independence. We sent astronauts to the moon on the strength of a national commitment; surely our energy problems rise to the level of a presidential crusade. Strong leadership is the antidote to our reliance on fossil fuels; it is time to seize this indisputably big idea—a turning point for wiser, sustainable use of our natural resources. Like a good stock portfolio, it is becoming increasingly clear that a diversified energy portfolio is a timely idea.

Cellulosic ethanol is an interesting case study. Unlike corn-based ethanol, planting, irrigation, fertilizers, or pesticides are not necessary to convert this raw material for fuel, although the grain crop may require such preparations. Moreover, no farmland is taken out of production because cellulosic ethanol can be made from waste discarded during the retrieval of grains. Wheat straw converted into biofuel reduces vehicle carbon dioxide emissions by 90 percent. There is good reason to advance these products so consumers can exercise their preferences in the marketplace.

Children born after World War II were often characterized as the "activist" generation. In fact, we were children
of enormous privilege whose parents sacrificed everything
so their kids could actively participate in American life.
Baby boomers were encouraged to enroll in scouting,
learn to play a musical instrument, join a Little League
baseball program, deliver newspapers on their bikes, and
thoroughly and safely explore every inch of turf in their
safe, expansive neighborhoods. We became student-body
officers, newspaper editors, debaters, cheerleaders, and
football heroes. Unlike their parents, ours signed up for
the PTA, and took the time to attend our games, recitals,
and after-school events. Our citizenship was shaped by our
eagerness to participate, to speak up, and to stand for
something.

Because World War II was the last war with unambigu-
ous unity of purpose, baby boomers had to figure out where
they stood on the Vietnam conflict. For this generation, the
environment became a unifying issue, something we could
all agree on. Clearly, from a boomer's perspective, the earth
needs our help. We recognize a call to action that affects the
future of life on this planet. We easily embrace a cause that
will make life better for our children and secure their future. To protect the next generation and beyond, baby
boomers are prepared to commit time, energy, and expertise. Like war, however, we must demand a complete and
decisive victory. Renewing the earth is surely one of the

greatest challenges this generation has confronted, and we understand how important it is to succeed.

We learned quickly that green is good, but we've been slow to learn that green is also hard. Going green in a serious way takes diligence, discipline, and patience. Alternative energy sources aren't perfect, cheap, or simple to implement. Recycling is a daily commitment; yet, our recycling efforts are frequently frustrated by items not yet accepted for recycling. We know recycling works, but we also know it could work so much better. The green revolution is not a fad. To be successful, it has to be sustainable; we have to reward and inspire human behavior to make it work. It's not enough to talk like true believers. We have to behave like true believers.

Although our missed opportunities are evident and abundant, there is also much to celebrate; in fact, some amazing benchmarks have been achieved. The U.S. Senate declared May 11, 2006, Endangered Species Day to encourage U.S. citizens to learn about endangered wildlife, success stories in species recovery, and opportunities to promote species conservation worldwide.

Indeed, the Endangered Species Act may be America's best environmental success story, although it has been subjected to severe criticism. As government, industry, and individual property owners have begun to find common ground, the enforcement of this act has been more effective. Furthermore, the act has suffered historically for lack of sufficient federal funding to enforce it. Many experts have concluded that the Endangered Species Act has led to important findings to support conservation action, and we may have bought additional time by its implementation. The following observation by W. R. Irvin is instructive:

> Given that the ESA [Endangered Species Act] is analogous to an emergency room which handles only the most dire cases, supporters . . . argue that it has been a success by getting hundreds of species off the operating table, though not yet out of the hospital.

A healthy Earth is defined, in part, by abundant and healthy populations of wildlife. A diversity of life in sufficient numbers to ensure their future is the goal of all conservation programs. The Endangered Species Act has been protecting wildlife since its inception in 1973. Critics contend the act is too restrictive on private landowners, but tinkering by the U.S. Fish and Wildlife Service, its principal enforcer, has rendered it more flexible. In addition, the U.S. Department of the Interior has expanded conservation tools available to private landowners and federal land managers. The Private Stewardship Grant Program and the Landowner Incentive Program assist landowners who voluntarily protect threatened, imperiled, and endangered species on their land.

We recognize that wildlife cannot effectively resist the advance of human habitat encroachment; living creatures have only the Endangered Species Act as a buffer against extinction. For that reason, we have supported its continuation against all challenges by private citizens, interest groups, and congressional opponents. In the past decade, no serious revision of the Endangered Species Act has yet provided the full range of protection its authors intended, so it remains largely unchanged. Endangered species such as the bald eagle, the gray wolf, and the American peregrine falcon have recovered under the protection of the act, so its achievements are notable and significant. Eleven species have been removed from the endangered species list because of their recovery since 1973, but only seven species have been removed from the list because of extinction. Independent scientific assessments estimate that 192 species might have been lost through extinction if the Endangered Species Act had not been in effect from 1973 to 1998.

Tough decisions have to be made to protect animals and our economy, so, whenever possible, compromises are sought. The science of conservation should always be tempered by the art of compromise as we seek truly win-win, entrepreneurial environmental solutions. The Department of the Interior's Safe Harbor program assures landowners will not be penalized if they make responsible decisions to protect endangered species on their property, an effort to reduce the adversarial moments generated by vigorous wildlife protection. To protect species, however, it is essential that we protect the legislation that continues to buffer wildlife from the real threat of extinction; we must extend the life of the Endangered Species Act through an admittedly delicate process of reauthorization.

Recently, the Bush administration proposed listing the polar bear as a threatened species under the Endangered Species Act. Officials from the Department of the Interior argued that receding sea ice may doom the polar bear to extinction within forty-five years. In this scenario, federal wildlife officials have assumed that Arctic sea ice is melting due to global climate change. Sea ice is expected to decline 50 to 100 percent by the end of the century. This intriguing development puts the Bush administration on record in acknowledging the deleterious effects of climate change. Currently, there are some 25,000 polar bears worldwide. A 21 percent drop in the polar bear population was recorded in Canada from 1997 to 2004 in part due to weight loss in females and concomitant changes in reproductive rates and cub mortality. The polar bear's situation symbolizes the difficult decisions that must be made, often based on limited data and uncertain projections. Many environmental groups that have pushed for an endangered species listing promote doomsday scenarios, when, in fact, no one can accurately predict the fate of polar bears—or any other species.

However, even if we take a cautious approach to the data, it would seem prudent to monitor the polar bear census. They are one of the earth's most charismatic species— beloved icons of Coca-Cola television commercials and appreciated by countless zoo-goers nationwide. It isn't necessary to link the receding ice to human activity to conclude that polar bears are in trouble if the ice disappears. We know that, in some locations, ice is receding; the facts are in, and no one disputes the phenomenon. However, whether this trend is precipitous or gradual, the time remaining to correct the problem is surely debatable. A cautious, conservative federal government is now ready to propose listing polar bears on the endangered species list. We can only hope that this initiative is a genuine example of badly needed presidential leadership and an indication of a growing consensus to protect the environment. Support for protecting endangered wildlife is widespread and bipartisan; every president and Congress must pay attention to the problem.

Threats to polar bear survival are not trivial, as extinction is looming for many forms of megafauna affected by habitat destruction. The journal *Nature* reports that human development in the Yangtze River area of China has ushered the rare freshwater dolphin to the brink of extinction in a short time. The paucity of dolphins located by scientists is too small to reproduce. This species has no other habitat in the world, so it may soon be identified as the first cetacean driven to extinction by human activity. A 2006 six-week survey in China by a Zurich-based conservation foundation (Baiji.org) did not find a single dolphin in the Yangtze River.

Fortunately, when wildlife populations reach the breaking point, people frequently figure out how to do the right thing. Recently, a committee of the United Nations International Maritime Organization authorized the narrowing of the shipping lane into and out of Boston Harbor, requiring vessels to make a wider turn around Cape Cod and shifting the lane about ten nautical miles to the north. The narrowing of the lanes was authorized to improve maritime safety and to reduce the likelihood of ships striking indigenous northern right whales whose movements are tracked by the Stellwagen Bank National Marine Sanctuary. When the new lane rules take, which is expected on July 1, 2007, the beneficial action on behalf of whales is expected to have no impact on commercial shipping interests, a virtual win-win for all concerned. Tragically, whale/ship collisions are common; of the fifty-two right whale deaths since 1986, 39 percent have been attributed to collision injuries. They also suffer from injuries inflicted by entanglements with fishing gear. The use of new fishing equipment with "breakaway" features should improve the right whales' chances of surviving such encounters.

Species conservation in America, and around the world, will always require a measure of common sense as communities seek to balance their interests in commerce with their quality of life. An enlarged community commitment to save wildlife and wildlife habitat has made it possible to bring alligators, bald eagles, peregrine falcons, and right whales back from extinction.

Conservation action always requires trust and compromise. Discord and disagreement among zoos, wildlife groups, and government agencies prevailed when the California condor captive conservation program was launched in the 1980s. Curators and scientists at the San Diego Zoo and the Los Angeles Zoo collaborated and eventually produced a healthy population of captive birds suitable for reintroduction into the wild. Bickering factions nearly doomed the program before it got started because some environmental groups apparently preferred extinction over the alternative: capturing and breeding the last remaining birds to preserve the gene pool. The captive-breeding strategy has been vindicated and California condors, the second-largest flying bird, can once again be seen flying magnificently in the wild.

To return endangered birds to the safety of nature, their habitat must be restored or relocated. Scavengers, such as condors, will encounter human refuse that is significantly more dangerous than the habitat of their ancestors, but they are helped by the ubiquity of recycling programs in the West. Thirty-five years ago, government recycling programs and private-sector recycling businesses were nonexistent. Today, recycling is a worldwide public/private enterprise sustained by the commitment of millions of individual citizens. Those born after the first Earth Day were brought up to recycle; some 3,000 recycling centers have been opened on Earth Day. We learned about recycling at preschool, practiced it at zoo camp, and wrote term papers about it in high school science classes. Today, the national average for states that recycle trash is 27 percent of the total material, while polls indicate that the number of people who indicate they have willingly participated in recycling programs is more than 80 percent. The universality of the recycling phenomenon should be regarded as a turning point in our struggle to revitalize the earth, and one of the most successful mass environmental actions in human history.

The enthusiasm of recyclers may be exceeded only by the zeal of antismoking advocates. Not long ago, people freely smoked on airplanes; even nonsmoking sections on airplanes were within reach of smokers. On long flights, passengers and flight attendants could not escape the smoke that permeated the entire cabin of a 747. America is well ahead of the rest of the world in freeing citizens from cigarette smoke, but other nations are following our lead. In these efforts, we have also set a standard for nonpartisan action. No single political philosophy generated these social changes. By consensus, smoking is now banned from many restaurants and taverns and almost all enclosed public places. Americans have been incredibly cooperative and compliant in supporting this major change in personal behavior. The engineering of smoke-free workplaces and the propagation of recycling communities demonstrates how much we can achieve when we rise above our differences and work together for the common good.

TALKING POINTS

1. Widespread scientific literacy is the key to generating sound solutions to longstanding environmental problems. Our best and brightest scientists must turn their focus to renewing the earth.

2. A climate of adversarial politics must be overcome if we are to reach a rational consensus on the environment. If we avoid the excesses of media hyperbole and insist on objectivity, we will build trust among citizens and experts, stimulate civil debate, and find common ground.

3. The Endangered Species Act is still working to protect wildlife from extinction. We regard it as a necessary buffer, an imperfect but significant factor in keeping extinction at bay and worthy of reauthorization.

4. A serious shift from fossil fuels to renewable alternatives deserves the attention of government and industry. Among our government's top energy priorities, a plan to significantly and rapidly reduce our dependence on oil should be vigorously pursued.

5
Partnering for the Earth

If we are together nothing is impossible.
If we are divided all will fail.

WINSTON CHURCHILL

WHEN WINSTON CHURCHILL said, "If we are to-gether nothing is impossible. If we are divided all will fail," the Allies were in the midst of the greatest military struggle of the twentieth century. Perhaps if Churchill were alive in our time, he would direct his supportive comments to one of the epic challenges of the new century, the pro-tection, restoration, and renewal of the natural world.

Our environment's current state represents both a unique challenge and a golden opportunity. If we respond with the ingenuity and diligence consistent with our na-tional heritage and our sense of duty, we will not only begin to resolve our environmental problems, but we will also launch an unprecedented epoch of economic prosperity. No person or entity, especially the business community, can afford to sit on the sidelines as our natural resources are squandered and degraded. The operating concept is remarkably simple: The business of conservation will gen-erate revenue, whereas a high price will be paid if we ignore the warning signs of our troubled planet. America will benefit economically and culturally from fostering part-nerships that generate new environmental business op-portunities. Working together, responsible environmental groups, neighborhoods, governments, small businesses, and major corporations will shape a future bound by a common cause—the environment—and against the com-mon foes of inertia, indifference, and apathy.

A widely held misrepresentation about business and conservation maintains that the two are necessary but largely incompatible. Fortunately, business partnerships on behalf of the environment are not some abstraction or propaganda ploy; they are well documented in scores of successful environmental projects. This chapter explores a few of them. If America is to be serious about leading a worldwide partnership, let us begin by tallying the need.

We must do as follows:

1. Lead the cleanup of the world's polluted lands and waters.
2. Invent new and clean sources and systems of energy production.
3. Lead the effort to reduce or eliminate greenhouse gases and airborne contaminants.
4. Assist nature in repairing our threatened reefs and wetlands.
5. Create a fully recycling society.
6. Recognize that we live in the environment—breathe it, eat it, and raise our families within it. Earth is our home.

As environmental enterprise expands to fill the need, the potential creation of prosperity is enormous. Endless tales of success can be used as guideposts. One leading environmental player is a uniquely entrepreneurial nongovernmental organization that promotes partnerships throughout the world; the highly regarded nonprofit

Conservation International (CI). A segment of their mission statement follows:

> to conserve the living natural heritage, our global biodiversity, and to demonstrate that human societies are able to live harmoniously with nature.

Conservation International's environmental profile has been elevated by its public record of success in establishing conservation partnerships with global corporations. For example, CI teamed up with the Ford Motor Company to establish the Center for Environmental Leadership in Business. Their joint goal is to engage the private sector worldwide to create solutions to critical environmental problems in which industry plays a defining role. The partnership works on both local and global levels to create best practices that support conservation and promote effective policy solutions. Their partnership takes on challenges in many of the world's most vulnerable "conservation hotspots," regions containing high levels of biodiversity. These locations are a global priority for conservation, and many contain a tremendous number of species at risk of extinction. In species-rich Ecuador, for example, CI has fostered a partnership in the northern coastal rain forest to preserve and connect the Choco-Darien Western Ecuador Hotspot. This hotspot is one of the world's most fragile regions of biological and cultural diversity. A small area, it is home to an astounding 800 species of birds and mammals, including jaguars and the rare mantled howler monkey. The situation is complicated because less than 2 percent of Ecuador's coastal rain forest is still reasonably intact. Like many places, rapid and devastating deforestation has ripped gaping holes in the natural fabric of this country's ecosystems. Ford; CI and its local partner, the Jatun Sacha Foundation (Ecuador's largest environmental organization); and the Climate Trust will reforest more than 19,000 hectares of land. In terms of global climate, the effort should have the effect of absorbing 9.5 million tons of carbon dioxide.

Aggressive reforestation provides habitats for species, beautiful destinations for ecotourists, new resources for nations dependent on a strong timber industry, and reductions of carbon dioxide from the atmosphere. To illustrate the latter point, scientists have calculated that deforestation from 1850 to 1990 released 122 billion metric tons of carbon into the atmosphere worldwide. The current rate is 1.6 billion metric tons per year, nearly three times the rate contributed by burning the fossil fuels of coal, oil, and gas. Arguably, deforestation is the world's most critical environmental issue. America and American companies can and should exercise global leadership through partnerships in reforestation. Strategic partners such as those established by CI ensure that industry has an opportunity to join in the efforts to protect the earth's fragile membranes and to do it with guidance from the world's best ecologists.

In the Netherlands, a unique partnership between Shell Oil Company and Dutch greenhouse businesses enables surplus carbon dioxide produced at the Shell refinery in Pernis, outside Rotterdam, to be pumped into greenhouses as an alternative to pumping it directly into the atmosphere as waste. The partnership was made possible because of an existing pipeline (built a decade ago but never used) activated to pump gas from Rotterdam to Amsterdam. Dutch entrepreneurs bought the pipeline from the government and then modified it for the distribution of carbon dioxide. The anticipated results are quite remarkable. Ninety-five million cubic meters of natural gas will be saved by the fruit, vegetable, and flower growers, and carbon dioxide emissions at the Shell plant will be reduced by 170,000 tons.

Investors are receiving 14 million Dutch guilders in tax relief in addition to a 4-million-guilder innovation grant from the Dutch government, so the incentives have really helped bring this innovation to fruition. The gas distributors make money, and the flower growers save money. Clearly, this innovative partnership has broken new ground in Europe and will be emulated by others. Once fully unleashed, robust government incentives combined with invigorated industrial creativity has the potential to transform the health of our global environment.

Belief in this entrepreneurial process is not sufficient; we must communicate, encourage, and nurture it. However, believing is a good place to start.

The Nature Conservancy has successfully brought business partners to the conservation table over the years. The membership base of this apolitical organization now exceeds 1 million, and the conservancy's reach extends into more than thirty countries. They describe their work as

a strong results-driven approach [that] is collaborative and pragmatic. The Conservancy actively partners with businesses of all kinds around the world to achieve lasting conservation goals, bringing a flexible and analytical approach to every project.

A good example of a Nature Conservancy partnership is the Sustainable Rivers Project operating on eleven American rivers, involving twenty-six dams flowing through thirteen states. In this case, the major partner is not a business (though many businesses are involved) but, rather, the U.S. Army Corps of Engineers. The Nature Conservancy/Corps partnership is actively improving dam management to protect the ecological health of rivers and surrounding habitat. At the same time, full involvement by other partners—those who need dams for flood control and to generate power—makes the process politically feasible. By cooperating on the Green River in Kentucky, the two partners found a more ecofriendly way to release water from the

Green River Dam. Their agreement to delay the release of reservoir water helped spawning by an array of local fish and mussel species. This change also benefited river recreation and associated industries because public access to the reservoir was extended more than one month.

The toxic impact of American industrial history is extensive but pales when compared with the polluted legacies of undemocratic regimes around the world. Communist and totalitarian dictatorships have systematically obscured the damage they have done, and in many cases, continue to do to the environment. As some of these countries shift toward becoming more prosperous democratic and free enterprise economies, they uncover the damage. America and other democracies have developed an array of environmental solutions. Fortunately, these can be brought to the attention of our emerging democratic neighbors. Private industries specializing in waste disposal are helping such countries to clean up their spills, but the pace of cleanup is slow. As totalitarian states are liberated into transparent democracies, the dirty industries of Eastern Europe can be equipped with modern standards of pollution control and clean energy technology. As they have converted to market economies—and not a moment too soon—Russia and other nations of the former Soviet empire are urgent candidates for green partnerships. Will America help? Will we export our environmentalism with the same zeal as we sell our software? We must, if we are to become leaders of the environmental economy in the twenty-first century.

The U.S. government has no shortage of agencies ready to partner with citizens, businesses, environmental groups, and other nations. The U.S. Department of the Interior is one of the most relevant. The Department of the Interior operates 57,000 facilities throughout the nation, including more than 800 dams and irrigation facilities that provide drinking water to 31 million people. A high-tech partnership between the Department of the Interior and the Navajo Nation provides off-grid energy, wireless Internet access, and e-commerce through entrepreneurial technology companies. The Navajo Nation benefits from telemedicine, satellite telephone links, and high-speed digital cellular connections. The partnership also improves access to water purification technology. The Navajo-Interior partnership is just one example of an extensive list of working relationships at the Department of the Interior, including collaborative efforts to preserve wildlife habitats while farming the land. The Department of the Interior's new and much-improved approach practices conservation through cooperation, communication, and consultation. This notion was explained in a recent speech by Lynn Scarlett, Assistant Secretary for Policy, Management, and Budget:

> The old environmentalism set prescriptive rules, requiring, for example, that ranchers conform to a four-inch stubble rule in which grass may not be grazed to shorter than four inches. Such rules impose

one-size-fits-all requirements that may have little relationship to ensuring healthy forage and ecosystems . . .
it led with the "stick," assuming that human motivation to excellence is best achieved through threat of
punishment rather than through incentives, example,
and inspiration.

Some of the best examples of entrepreneurial partnerships are found in ocean conservation circles, with sea
turtle conservation as perhaps the most useful model. In
Florida, a group of environmental organizations, businesses of all stripes, and government agencies has been
working together to recruit private homeowners and
homeowner associations to participate in sea turtle conservation. The program is called NESTS, or Neighbors Ensuring Sea Turtle Survival. People who live near sea turtle
habitat are asked to save sea turtles "one nest at a time."
Participation is certified at the level of "partner, guardian,
and champion," depending on how much each person
accomplishes. Examples of actions that might be taken
include an organized effort to turn down beach lighting
that may disturb nesting or the emergence of hatchlings.

Fifteen organizations are currently active in the NESTS
program along with the coordinating institutions Ca-
ribbean Conservation, the Ocean Conservancy, and the
for-profit Walt Disney World. The effort is also partially
funded by proceeds earned from the Florida Sea Turtle
License Plate program. The ubiquitous, highly visible nest-

ing habits of sea turtles have made a lasting impression on Florida's businesses, citizens, and visitors. Employees of Florida beach hotels enthusiastically (and diplomatically) enforce lights-out regulations. When hotel guests are asked to turn off lights in their rooms, they comply and become willing participants in the sea turtle conservation drama. Although everyone now sees the value of sea turtle conservation, it took flexible, incentive-based conservation policies to make it happen.

In Costa Rica, sea turtle protection has become a highly personal mission for Laura Jaen, who has organized local women to educate others about sea turtle conservation. With guidance from CI and the Leatherback Trust, the women (many of them local tour guides) raise money from the tourists who visit the nearby Las Baulas National Marine Park. The local beach at Playa Grande is incredibly beautiful, but it is also one of the last thirty or so nesting grounds for the leatherback, the world's largest sea turtle. Scientists have calculated the leatherback's catastrophic decline to a level that is 90 percent less than their historic population. With dedicated women like Laura Jaen in the field, a comeback is challenging but not impossible. Around the world, sea turtles have been carelessly captured in fishing nets and in longlines, while nest sites have been destroyed because development standards failed to consider the impact of construction on nesting behavior.

As a charismatic flagship species, sea turtles are easy to love. At the Living Seas exhibit at Epcot in Orlando, Florida, an animated sea turtle is the star of a new conservation-based program. The critter named Crush first appeared in the Disney animated movie *Finding Nemo* (2003). Amazing technology drives this interactive show, and media exposure has helped to elevate sea turtles to priority status, inspiring a growing number of advocates who have championed their protection.

While the plight of some sea turtle species has improved, others remain in danger. Yet, it is essential that the good news of conservation successes be disseminated to counteract the bad news that dominates the headlines of newspapers, magazines, radio, and television. For example, the Kemp's ridley sea turtle was as close to extinction as a species can get. At one time, nesting females numbered in the tens to hundreds of thousands, then declined to a few hundred individuals. Through the implementation of environmental partnerships, the numbers have risen into the thousands again.

One way to protect species is to concentrate our efforts on the protection of biodiversity hotspots. To qualify as a hotspot, according to experts at CI, a region must meet two strict criteria:

- At least 1,500 species of vascular plants must be endemic, meaning they are found nowhere else on Earth.
- A hotspot must have lost at least 70 percent of its original habitat.

Conservation International has identified thirty-four hotspots in its most recent analysis. The degree of lost habitat can be expressed by comparing the 15.7 percent of original habitat with the remaining 2.3 percent of Earth's surface now occupied by the thirty-four hotspots, a precipitous loss of 86 percent. Among them, they contain at least 150,000 plant species as endemics, 50 percent of the world's total. The 11,980 estimated terrestrial vertebrates endemic to the known hotspots represent 42 percent of all terrestrial vertebrate species in the world. When you add terrestrial vertebrates found elsewhere, the figure jumps to 77 percent. Hotspots are profoundly rich locales for the world's wildlife, and the amount of biodiversity in hotspots is extremely high, so these remarkable places represent an urgent priority for conservation. If we are not successful in protecting them, we could lose more than half of the world's wildlife.

By protecting wildlife, especially undiscovered species, we also protect our opportunity to discover valuable new organic material. Thomas Eisner of Cornell University defined *chemical prospecting* as the exploration of nature to discover new natural substances potentially valuable to humanity. Eisner rightly considered the natural world a vast chemical treasury, but a source of wealth that is diminished by losses in biodiversity occurring throughout the world at a disturbing pace. Eisner envisioned opportunities for developed countries to partner with biodiversity-rich underdeveloped countries to the benefit of both. Such a partnership was formed in 1991 between the pharmaceutical company Merck and Costa Rica's Instituto Nacional de Biodiversidad. Merck agreed to pay royalties for any products developed as a result of their exploration agreement and further provided a $1 million cash advance to their Costa Rican partner.

A successor to the partnership, the International Cooperative Biodiversity Groups Program was established two years later and funded through the National Institutes of Health, the National Science Foundation, the Agency for International Development, and the U.S. Department of Agriculture. As many as twelve countries with tropical resources have participated in partnership with universities, botanical gardens, museums, conservation organizations, and pharmaceutical companies. This program was more transparent by providing public access to their research. Compounds and substances not immediately developed or tested by partnering entities can be used theoretically by others who examine the data. By easing up on proprietary controls, product development can be expedited and the wealth can be spread among a greater array of partners. In a recent review of the status of chemical prospecting, Eisner lamented that political difficulties have dampened enthusiasm for exploration as well as essential conservation strategies linked to economic development. Adding to the problem is an unfortunate shift in the chemistry mainstream that has caused the field of natural products chemistry to be systematically downsized in both industry and academia. According to Eisner,

> chemical prospecting will doubtless suffer in the interim, as will conservation programs slated to be linked to the prospecting effort. One can only hope that logic will prevail and that natural products chemistry, modernized perhaps by closer affiliation with molecular biology, will eventually be readmitted to center stage.

The state of chemical prospecting today, still full of potential and opportunity, is symptomatic of many advances in entrepreneurial environmentalism. Often the new approach is displaced by another more promising or cheaper technology. The playing field is by nature mercurial and finicky. Further, the opportunity to commercialize nature is not acceptable to everyone. Although a new U.S. National Park Service plan calls for revenue sharing with companies that discover and market organisms within the national park system, not-for-profit environmental groups have protested and sued to restrain the partnership. Governments and private foundations will have to continue to extol the marketplace of environmental ideas so that we don't abandon a good idea too early or interrupt its development with bickering about who gets paid. We are reminded of hungry lions that encounter a million wildebeest. Which one shall they eat first? The sheer magnitude of the opportunities ahead can paralyze and stifle innovation if we don't find a way to get on with it.

Costa Rica is acknowledged as a model for protecting its natural wealth in an ecosystem-based economic strategy. As an ecosystem degrades, the "services" it provides for humanity, such as fresh water and fertile soil, also decline. Developing nations are especially affected because they are tempted to overexploit their resources for quick-fix economic gain. Moreover, the intrinsic value of the natural world is often ignored or underestimated. For decades, CI has worked in Costa Rica to change this paradigm. The fundamental problem for the government and its nonprofit environmental partner was to conserve ecosystems without damaging the prospects of local workers.

Conservation International professionals responded by organizing a partnership uniting the governments of Panama and Costa Rica to combine ecosystem protection with sustainable development. By recruiting funding from McDonald's Corporation, this project pioneered cooperation between global business interests and conservation. Rural farmers received direct support from the partnership, which enabled them to refertilize depleted soil, manage water resources by protecting adjacent forests, and start an innovative and sustainable "conservation coffee" agroforestry. While farming thrived, widespread protection of the rain forest enabled ecotourism to emerge; it is now the nation's most important revenue-producing industry. Government incentives and outside investment sustain the gift of biodiversity, fuel the ecosystem that collects water and absorbs greenhouse gases, and protect the natural beauty of a region that is attracting tourists in record numbers.

In Costa Rica, farmers receive financial incentives for ecologically sound practices. Former cattle pastures are reverting to rain forest, as Costa Rica has become the first developing country to reverse deforestation. From a low of 21 percent in 1980, tropical forests now cover more than half the country, which has turned Costa Rica into a Central American economic powerhouse. Costa Rican officials are in demand as they teach other nations to create a conservation economy. This is a model worth emulating.

Soy products, like coffee, can be produced as an environment- friendly commodity. According to the World Wildlife Fund, expanding soybean cultivation has damaged forest and savannah ecology in South America, but new methods protect primary vegetation. In 2006, Fenaco, a Swiss company, imported 1,000 tons of soy pellets produced in an ecofriendly manner, following the Basel Criteria for Responsible Soy Production, which state that producers must refrain from clearing virgin habitat and instead grow soy on existing cattle pastures. The growers must also maintain soil and water quality. Nearly one-third of all soy fed to domestic animals in Switzerland now meets the Basel Criteria. The power of free enterprise suggests that European consumers will prefer and demand environment-friendly soy, which motivates South American growers to comply with the stricter Basel Criteria standards. The Basel Criteria also call for social changes in the workforce, such as banning child labor in the fields. If the new model for soy production remains cost-effective, the Swiss experiment looks promising.

Ecotourism exposes citizens to the reality of environmental degradation, the value of pristine habitat, and the opportunities to support conservation. Although the cruise industry provides human access to beautiful, clean beaches and ocean vistas, the same industry can potentially harm the untarnished isolation we crave through wilderness ecotourism. Too many visitors can cause harm to these natural places, defeating the purpose of ecotourism. Cruise companies must maintain a delicate balance. Likewise, local people who work in the industry must temper their aspirations. Poorly managed ecotours, wherein tourists overrun environmentally sensitive areas, can also adversely affect the local economy, whereas well-managed tours can bolster the economy. Two-thirds of the world's ecotourism cruise destinations are found within biodiversity hotspots; therefore, CI has been active in organizing cruise lines to step softly from ship to shore.

Cruise lines are not always a threat, however. Many offer opportunities to support and promote conservation. In partnership with local governments and shoreline operators, they can raise the awareness of their passengers so ecotourists learn how to behave responsibly in the ports they visit. Local cultural and environmental issues can be shared onboard, and tourists can be directed to conservation and social programs that need help. Nothing compares to seeing nature at its best, and the cruise lines can help to make conservation-minded passengers even more committed to the environment. The net result often is a partnership that provides local economic incentives to protect wildlife and wilderness sites in order to accommodate an ecofriendly clientele.

In the realm of environmental education aboard a cruise, no one does it better than Disney. In Grand Cayman, Disney Cruise Line is working with tourism officials to encourage children to learn about the environment. Disney's *environmentality* concept is promoted by the animated figure Jiminy Cricket, who helps to communicate the need to learn more and take action locally. Each season, some 1,500 local children work together on a class project in which they learn to care for the environment by working in their community. The program is under way in many places that Disney now operates, including California, central Florida, and Hong Kong.

Disney's environmental education programs are visible in their theme parks, and the cruise line reinforces the conservation theme throughout the ship. Onboard recycling is practiced. Disney technicians have studied in detail the opportunities to save energy, recycle resources, and reduce waste disposal on the ship. During the cruise, guests can converse with the ship's dedicated environmental officers to learn about efforts to protect ocean habitat and the creatures that live there.

Carnival Cruise Lines is working with a group known as the International SeaKeepers Society to gather real-time data on ocean water quality. Scientists at the University of Miami monitor levels of ocean pollution and changes in global climate detectable in the ocean. Data gathered at sea is transmitted by satellites and used by university scientists and other collaborators in government and industry. In this way, Carnival contributes to environmental quality control on behalf of the tourists they carry to exotic destinations around the world. Every cruise ship in the world could participate in ocean monitoring programs such as this one. The cruise line passenger population exceeded 11 million in 2005; clearly, cruise lines can be major contributors to mainstream conservation.

We recommend that consumers use their marketplace knowledge to select cruise lines that advertise their corporate commitment to protect the environment in the venues consumers frequent. Cruise ships such as the *Disney Wonder* are steadily and significantly reducing their environmental impact with the creativity of a Disney Imagineer. Indeed, the company has called on its environmental expertise to study the many ways that a ship can cruise with scarcely an environmental ripple. No doomsday environmentalism can be found aboard the *Wonder*. Instead, joyful, inspirational, optimistic environmentalism characterizes the Walt Disney Company's commitment to efficient and sustainable operating practices.

The experience of each successful partnership inspires others to make a difference. It will surely take a world of differences to renew the earth, but we've made an encouraging start just a few years into a new environmental epoch. Awareness, understanding, and consensus are needed to mobilize earth-friendly entrepreneurs. Given what we know, it is time to build on the partnerships that are already working. By sharing the story of their successes, we combat the dangerous and incorrect perception that the environmental struggle is unwinnable.

With so many examples of successful environmental partnerships, the world should be well on its way to full recovery. However, the depth and breadth of planetary problems preclude a limited, quick fix. A more extensive network of collaboration will be necessary, and we must all be accountable to make our individual contributions to the cause. To prevail, we must all heed Churchill's call to unity, a message that resonates in our time.

TALKING POINTS

1. Hardly adversaries, businesses and conservation organizations have formed compatible partnerships to achieve win-win solutions in the real world.

2. Experienced nongovernmental organizations, such as Conservation International and the Nature Conservancy, teach others how to form partnerships in devel-

oping countries where diversity hotspots require swift action and local people need meaningful employment.

3. One of the most enlightening conservation partnerships is saving sea turtles and sea turtle habitat around the world. These are models that should be implemented with other endangered populations.

4. Green enterprise is becoming the norm throughout the world. Scores of new private-public environmental partnerships are showing up on the Internet daily.

6

Entrepreneurial Environmentalism

The living world is dying; the natural economy is crumbling beneath our busy feet—Science and technology led us into this bottleneck. Now science and technology must help us find our way through and out.

EDWARD O. WILSON

AS YOU DRIVE SOUTH ON the Florida turnpike, antici-pating the magical properties of Everglades National Park and the Florida Keys, you are startled by massive mounds of trash heaped much too close to this modern roadway—higher than you would ever expect a landfill to reach. Above the heap, hundreds of shorebirds circle, punctuated by the dark silhouettes of the ubiquitous black vultures so common to this region.

Something is wrong with this picture.

How did this happen and what can we do about it? Neither speed nor insulation in the most modern vehicle will keep the foul, inescapable odor at bay. Landfills are proliferating state by state at a rapid pace. Some are visible from the roadside, while others are tucked away beneath our personal radar. Whether landfills are a ticking time bomb or a medical emergency the challenge they present leaves a provocative and lasting image.

But does this image contain the power to trigger action by a citizen, a state, or the nation? Are we smart and de-termined enough to solve such a massive and growing problem?

In a fascinating example of innovation, entrepreneurs and governments have discovered a good use for munici-pal solid-waste landfills, in which landfill gases are con-verted into a type of green energy encouraged by gov-ernment subsidies and tax credits that are attractive to investors. By collecting and dispersing landfill gas for home heating and electricity, green energy investors expect to operate a profitable business at the landfill site.

Landfills are commonplace in America; more than 2,000 municipal sites have been registered. Currently, our citizens generate some 230 million tons of solid waste annually. About half of this tonnage ends up in landfills. Landfill gas contains mostly methane and carbon dioxide, which are greenhouse gases, so there is good reason to try to limit or to exploit these emissions. The Environmental Protection Agency promotes the use of landfill gas as a renewable, green energy source and has established partnerships around the country with its Landfill Methane Outreach Program. An example is the Salt Lake City, Utah, area energy partnership. The partners—a coalition of federal, state, and local governments; power companies; and entrepreneurial alternative energy firms—expect to provide power to more than 2,500 homes in Murray City, Utah. The project will generate three megawatts of clean power, a good start for this region.

A similar program operates at the foot of the Black Mountain Range in western North Carolina. The nonprofit corporation EnergyXchange burns landfill gas as an energy source to support small enterprise in crafts and horticulture and local homeowners. The six-acre landfill is situated next to the community's highest-energy demand facilities, while the organization is led by a group of local government and business leaders who are strongly committed to principled environmental stewardship. Programs like this are sprouting up everywhere, providing a small but meaningful boost for communities that are seeking to make a difference. As these programs expand and take on a greater proportion of our power requirements, landfills could serve a useful purpose during their lifetimes, turning a former pariah into a community asset. As valuable as this form of energy may be, it is the redirection of greenhouse emissions that is most important to the technology. Without it, the gases simply escape into the atmosphere where they do their damage.

New technology is also taking shape in St. Lucie County, Florida, as local government has opted to vaporize garbage at temperatures of 10,000 degrees Fahrenheit, as hot as the surface of the sun. The $425-million facility will turn trash into gas and rock by-products from the application of plasma arcs. By 2008, the county expects to use this method to vaporize 3,000 tons of garbage per day. The entire landfill, now a mass of 4.3 million tons, should be completely eliminated in two decades. Geoplasma, an Atlanta-based company, foots the bill for the plant and has promised to use the by-products, known as "slag," as a marketable commodity used in constructing roads and highways. The gas produced at the site will run turbines, creating some 120 megawatts of electricity sold to the grid. The facility will operate year round, nonstop, on one-third of the power it generates. About 80,000 pounds of steam will be produced daily, which has been promised to a local juice company. Some of the production materials, esti-

mated at 600 tons per day, will be sold for road construction. Geoplasma's solution exploits the inherent power
of the landfill as it eliminates, or at least slows, landfill expansion. Louis Circeo, a Georgia Tech scientist,
has suggested that large plasma facilities deployed nationwide could theoretically generate electricity equivalent to
twenty-five nuclear power plants. This appears to be a significant opportunity. In 2003, Americans generated 236
million tons of trash, with an estimated 130 million tons
dumped directly into landfills. Although critics worry
about the level of sulfur dioxide and particulate emissions
from plasma-arc gasification, similar technology operating in Japan has met much stricter emissions standards
than required in the United States.

In the Texas cattle town of Hereford, Panda Ethanol,
Inc., is building an ethanol factory stoked by mountains of
local cow manure. The company's CEO labeled his project
the "Saudi Arabia of manure," which has the potential
to produce 100 million gallons of ethanol annually. This
would amount to the gasification of 1 billion pounds
of manure. By using a portion of its natural resource
to power the company, Panda Ethanol will save nearly
365,000 barrels of oil in annual operating costs. Some
ethanol plants use electricity from coal, a ubiquitous fossil
fuel, but Panda Ethanol provides more than 90 percent of
its own energy needs.

How does the "poop-to-pump" approach work? The
plant heats the manure until it releases methane, which it

burns to produce the steam that fuels the plant. In the end, the highly destructive methane is incinerated. Furthermore, the use of manure reduces the risk of water pollution produced by farm runoff. The ash by-product can be used to make cow bedding and cement. With aspirations of expansion and public investment, Panda Ethanol is positioned to meet the growing demand for alternative sources of energy. Such projects are not inexpensive as the company secured $160 million to finance the venture. Clearly, investors have recognized its potential to generate profit while helping to protect the earth.

Landfills increasingly consist of discarded technology: computers, mobile telephones, and other electronic devices. In response, some companies, such as Dell, Hewlett-Packard, and Nokia, voluntarily take back some of their products consumers wish to discard. However, many firms have not yet adopted this recycling policy. Experts estimate that "e-waste" has reached 58,000 tons, and it is growing by more than 10 percent each year. Because many electronic devices contain hazardous materials, it is dangerous to discard them. Other forms of e-waste are refrigerators, televisions, and air conditioners. The desire to own fast-developing technology hastens the growth of e-waste and shortens the useful life of these products. Some companies, such as Fuji Xerox in Thailand, operate recycling plants for its products. Globally, e-waste is calculated as a 20- to 50-million-ton problem. Recycling is essential, but the design and production of greener, safer devices is a better solution. Innovations can be aggressively marketed, and it is likely the informed buyer will prefer the greener product. Still, a clever entrepreneur is going to figure out how to turn this problem into a 20- to 50-million-ton business opportunity.

Landfill technologies are sometimes criticized because of their potential competitive advantage over proactive resource reduction and recycling methods. For example, government must be careful not to provide tax benefits for landfill technologies at the expense of recycling plants or they may put recyclers out of business. Recycling plants need to be competitive. Most experts prefer to avoid creating future landfills by relying on aggressive resource reduction and recycling programs. Pitting landfills against recycling is a mistake. An analysis by the National Resources Defense Council first recommends avoiding increases in landfill biomass by an emphasis on reduction and recycling. The council also recommends burning landfill gases to reduce toxicity and greenhouse gas emissions and using landfill gases in energy production. The council's analysis confirms the value and potential effect of emerging landfill technologies.

Creative solutions to save wildlife habitat may involve a reversal of technology. One example is a partnership by the Nature Conservancy and Environmental Defense Fund (EDF), which advocates market-based solutions to environmental problems. They have entered into collaboration to eliminate bottom trawling by commercial fisherman operating on the West Coast. The target area covers 6,000 miles of ocean no-trawl zones. Local trawlers have agreed to participate in exchange for the Nature Conservancy's willingness to buy them out of the trawling business, including the fair-market purchase of their boats. In addition, entrepreneurial conservationists are helping local fishermen to purchase smaller vessels suitable for more selective and sustainable methods of fishing. The old trawling technology was designed to scoop truckloads of fish so effectively that it endangered the supply. This win-win idea is supported by all parties and has united factions long polarized by endless debate. The genius in this solution is the use of private money to offset the social and economic costs of marine conservation, while preserving livelihoods and restoring fish populations. Moreover, traditional fishermen have changed their attitudes.

By careful scientific analysis, the no-trawl area has been projected to protect two-thirds of the biodiversity along the specified ocean shelf. If the methodology is feasible elsewhere, the fishing industry could become sustainable throughout the West by the more appropriate technology of hooking and trapping. Efforts of fishermen and conservationists in California demonstrate that low-volume, high-value fishing can work and also appeals to consumers who care about how fish are caught. In this case, innovation has resulted in a return to simpler, less devastating forms of fishing technology.

Such examples indicate that we have entered into a new

period of compromise and collaboration after years of contentious confrontation. Industry and government are reaching out to environmental groups that are willing to cooperate rather than criticize. Brokers such as Conservation International and others are diligently working to access business and harness their economic power for the common good. Nongovernmental organizations are driving these conversations because of the NGO's inherent flexibility and speed. Proactive collaboration is further illustrated in the case of Tiffany & Company, which joined forces with Earthworks to investigate environmentally sound methods for mining gold and silver. As a result of this dialogue, a collection of best-mining practices was established for an industry that wanted to increase public confidence. Some may say these are uneasy alliances, but it is likely that the inclusive dialogue on creative environmental problem solving will continue to thrive.

Municipalities throughout the nation continue to take small but important steps. No city has been more active than Portland, Oregon, which coalesced around green action decades ago. Portland's greenhouse gas emissions have receded to 1990 levels compared with a national increase of some 16 percent. By all accounts, the measures taken by local industry and government have actually improved the economy. City fathers advocated conversion from diesel to biodiesel, and they have legislated a 10 percent ethanol requirement for other vehicles. The city will also implement wind and solar power for government vehicles by 2010, and they have altered streetlamps to use low-power bulbs. For years, Portland citizens have used bicycle trails and an expanded mass transit system to discourage the use of automobiles. This municipal model demonstrates that innovation doesn't have to be costly or complicated. Moreover, Portland's green initiatives illustrate that American cities can start small to improve the environment in their communities. In an interesting footnote, Portland citizens have expressed considerable pride at their growing reputation as America's greenest city.

Many other American cities are encouraging their citizens to change their behavior and adopt greener practices. Albuquerque, Los Angeles, New Haven, and Salt Lake City, for example, offer free parking for green cars. These and other incentives are motivated, in part, by clean-air goals adopted in the "U.S. Mayor's Climate Protection Agreement."

High-occupancy-vehicle lane preferences have also been offered to green automobile owners, and the local government of Wilmette, Illinois, has offered a two-thirds discount on vehicle registration to drivers of hybrid motor vehicles. Local governments have learned that environmental incentives are popular with the voting public, and they see an opportunity to offer environmental leadership despite apathy at the federal level. Corporate incentives are also trending green. Bank of America announced it would provide reduced mortgage rates in 2007 for homes that meet certain energy-efficiency standards.

Going green has been adopted by the entertainment industries and other prominent celebrities. Even the enterprises of professional sports have done their part. At the Super Bowl, the National Football League routinely plants seedlings to offset greenhouse gas emissions produced by the event. This practice was initiated at the thirty-ninth Super Bowl in Jacksonville, Florida. The NFL worked in cooperation with Oak Ridge National Laboratory and Princeton University to calculate the amount of carbon dioxide produced by Super Bowl events and matched it with the number and type of tree seedlings needed to absorb an equal amount of carbon dioxide.

Similarly, the 2006 World Cup soccer championship was also planned to operate as a carbon-neutral event. World Cup organizers met their objective by investing in a project that led to positive avoidance of emissions greater than or equal to the amount generated by the World Cup event. By replacing coal-fired boilers for the African Realty Trust in South Africa, World Cup officials enabled the company to burn sawdust from surrounding sawmills. This change minimized waste, reduced greenhouse gas and soot emissions, and fostered sustainable local development. However, some caution is advised, as some of these efforts may be regarded as little more than public relations devices. Critics have charged that individuals or companies that invest in such offsets without changing their own behavior are not committed to environmental reform.

In a spin-off from the Super Bowl extravaganza, and a

boost for cleaner fuels, General Motors provided hybrid-powered buses to transport journalists and VIPs at the fortieth Super Bowl in Detroit. The five buses in service saved more than 100 gallons of fuel compared with conventional buses. They also produce up to 50 percent fewer nitrogen oxides and 90 percent fewer particulate, hydrocarbon, and carbon monoxide emissions compared with diesel buses. In their press kits, GM marketers identified 380 GM hybrid-equipped buses operating in twenty-nine U.S. and Canadian cities, with another 216 scheduled for delivery in 2006. Big events like the Super Bowl represent opportunities for host cities to promote new technologies. To sustain Earth initiatives, people must be led to believe in their value. With business executives such as the CEO of Wal-Mart proclaiming a company goal of a 100 percent renewable supply of energy, no waste production, and products that sustain the environment, citizen and corporate commitment is bound to grow.

Wind turbines represent one of the most fascinating new energy technologies, and Florida Power and Light is the largest owner of wind farms in the United States. Florida Power and Light (which is actually a subsidiary of FPL Energy) owns forty-four wind farms in fifteen states, producing 3,210 megawatts of electricity, which is enough to provide power to 1 million homes. Wind power now accounts for about 25 percent of FPL Energy's generation portfolio. Because of tax breaks, wind power is also highly profitable, but the Production Tax Credit legislation has been inconsistently applied by Congress, and the tax credit is set to expire again in 2007. To continue to develop this unique source of power, the least that can be done is to extend the credit for a longer period, for example, five to seven years as some have recommended. The longer a tax credit remains on the books, the more significant its financial impact. As expanding wind farms require the acquisition of dedicated land, there are detrimental effects on environmental aesthetics, including noise pollution. However, cheap and profitable energy supplied by wind technology appears to be here to stay.

A way to encourage the development of wind power has been promoted by FPL flyers in Florida newspapers. The company actively markets its Sunshine Energy program to Florida citizens. For a small fee—less than $10 a month— Floridians can support wind power research and development. As the percentage of alternative energy is developed, customers will receive some of the benefits at home; however, if they don't receive it themselves, other customers will receive a small percentage courtesy of those who are willing to share. In essence, FPL customers can become stakeholders in the growth of alternative energy throughout the nation. Soon, Florida companies will be encouraged to participate, augmenting the investment of individuals. Because customers/investors don't expect to benefit directly right away, this program may be a true example of environmental altruism. Green Energy, a similar program, is operated by Georgia Power at the Seminole Road Landfill in Dekalb County, east of Atlanta. The facility produces methane gas from decomposing municipal waste to power electric generators and sells green energy to business and residential customers.

Sources of clean energy may also be found in the world's oceans. Scientists at Florida Atlantic University, in collaboration with the Ocean Renewable Power Company, are placing a turbine the size of a tractor-trailer into the Gulf Stream to learn how much energy can be produced by tapping this warm-water source. This technology operates much like a land-based windmill, floating underwater anchored to the seabed and connected to a land-based utility company. It is an environment-friendly energy system with no fuel, gas, or oil discharges that could harm sea life. If the energy module performs as expected, a sufficient number of turbines could power small cities along the coast. Collaboration among university-based scientists, entrepreneurs, power companies, and governments are yielding promising technologies every day.

Of course, big ideas attract new investors. Venture capitalists from California announced on February 16, 2006, their commitment of a considerable sum on the promise of green technologies. Kleiner, Perkins, Caufield & Byers are investing $600 million in biofuel and solar energy over a three-year period. Their recent scouting tour of Brazil demonstrated the efficacy of ethanol, where 100,000 tons of raw sugar is converted into 21 million gallons of ethanol every year. Soon, most of Brazil's automobiles will be powered by clean-burning ethanol derived from sugar. In its press release, John Doerr, a leading partner in the firm, stated that the global transition to clean forms of energy is the "biggest business transformation of the decade, and maybe even the first half of the century."

Free enterprise is regarded by many environmental entrepreneurs as the key to a better world. For example, Stanford ecologist Gretchen Daily has been involved in launching the world's first clearinghouse for ecosystem assets. To offset the effect of development, such a marketplace encourages developers to "buy" species and support investment in the conservation of habitat. Investors who want to develop in sensitive areas make species investments to offset the effect of the development. In the case of endangered woodpeckers, for example, Professor Daily observed:

> Development is just marching on, despite the Endangered Species Act. And these flexible, voluntary schemes are much more likely to be followed than the sledgehammer approach. It's better for the woodpecker because something is being done . . . You don't buy title to the land, you buy the woodpecker on it.

Another example of a pervasive attitude shift in marketplace behavior is the change in government purchasing habits. For example, former New York Gov. George Pataki issued executive orders asserting that all public schools must purchase only "environmentally friendly" cleaning solvents, while the neighboring state of Massachusetts spent $145 million on environmentally preferred products in the 2004 fiscal year. Even government computer purchases were subjected to human health and environmental considerations in Massachusetts according to a report in the *New York Times*.

In King County, Washington, and throughout the nation, government police vehicles are running on alternative fuels, and hybrid vehicles are being selected for mass-transit systems. At the federal level, executive orders on purchasing have required more spending on recycled products. The facilities of the Environmental Protection Agency are powered by alternative forms of energy, which sets a good example for other agencies. The standards of the EPA, which are ahead of the environmental curve, are being adopted at the Department of Energy and the Department of Defense, among others. For example, the U.S. Army is now using recycled paper or paper with fewer harsh chemicals within its huge paper budget of $100 million per year.

Incredibly, these standards now extend to weaponry. A recent nuclear submarine purchase by the U.S. Navy was accompanied by orders that the ship use only propellants, paints, and cleaning fluids that would reduce harm to the environment. Perhaps the most interesting trend is the use of lead-free bullets on military firing ranges, an effort to reduce the effects of ground pollution. These government practices are essentially nonpartisan, representing another example of agreement and consensus on environmental issues.

The marketplace supports green technology abroad. Some of the most active environmental entrepreneurs are at work in India, Japan, Singapore, Taiwan, and China. Hundreds, even thousands, of promising ideas can be found in annual reports, technical reports, Internet and print media accounts, web-logs, conference proceedings, textbooks, and market hyperbole produced to woo investors. Sifting through this mass of material, the reader is impressed by its diversity and the enthusiasm behind it. This is indeed an exciting time to live in America, an incredibly free country where big ideas can be developed by anyone clever enough to recruit the resources and patient enough to act on a plan or a dream. Our expertise is valued and imitated by other nations, and we freely share much of it with the international community. Synergistic free enterprise is the key to green development in impoverished nations.

We opened this chapter with a quote from E. O. Wilson, one of the world's leading biologists. Wilson has argued that the world's "natural capital" (arable land, groundwater, forests, marine fisheries, and petroleum) are finite resources. Overharvesting and habitat destruction degrade the value of these assets. Wilson warns that at present rates of habitat destruction, half the world's plant and animal species could be lost by the end of the century. This catastrophic loss of biodiversity would be a severe blow to the natural economy. Greener human habits will act to slow the pace of loss, but government and industry will have to be ever more entrepreneurial to turn the tide and renew the earth. A commitment to innovation will be the key to our success. Further, innovation on a grand scale is not inexpensive. Government can stimulate innovation, but the private sector will lead the way to innovative growth and development.

The chapter that follows takes up a different road to renewing the earth; a road paved by passionate philanthropy. The fate of people, wildlife, and habitat is linked in important ways. One significant challenge may be the integration of a comprehensive environmental policy that provides equitable benefits for ecosystems and civilizations alike. As Professor Wilson observed:

> The greatest challenge of this century will be how
> to raise the lives of people everywhere to a decent level
> while keeping as much of the natural environment as
> possible intact. The two goals can be approached in a
> synergistic way, in which progress in one enhances
> progress in the other.

The challenge that Wilson articulates is being played out today in New Orleans, Louisiana, in the aftermath of Hurricane Katrina. In this devastated city, where people have lived for decades below sea level, visionary city planners want to restore wetlands and provide for a more compatible mix of people and indigenous ecosystems. It will take a heap of human ingenuity combined with altruism and leaps of great faith to craft the win-win that is needed on the Gulf Coast. Political polarization has not helped the citizens of New Orleans to recover, and it will surely impede progress toward a consensus as communities with decades of history and a rich cultural heritage grapple with the consequences of this natural disaster. Somehow, the citizens of New Orleans will have to pull together and find a way to implement some form of green development strategy that is conducive to reconstruction in the fragile, delta ecology. Appropriate development and a modern, fail-safe levee system is necessary to facilitate quality of life and restore the vitality of the once and future city of New Orleans.

TALKING POINTS

1. Innovations in landfill technology are making it possible to use landfill gases to provide inexpensive forms of energy, ultimately eliminating landfills, a prime example of entrepreneurial environmentalism at work.

2. Local government, from the state to the municipality, is flexible enough to lead on environmental issues, demonstrating that conservation is best served when we act quickly and locally.

3. Wind technology is gaining traction as an alternative

energy source, but federal incentives must be renewed so energy companies will continue to invest in wind turbines in the long term.

4. Environmental protection and higher living standards must be given equal emphasis to ensure that people embrace and don't oppose environmental initiatives.

7
Environmental Philanthropy

Do what you can, with what you have,
where you are.

THEODORE ROOSEVELT

THE AMERICAN PRESIDENT known far and wide as "Teddy" was a passionate environmentalist/conservationist and a genuine pioneer, who reflected the prevailing spirit and attitudes of the period. He passionately hunted and fished throughout America and abroad. His larder from an African safari in 1909 was impressive by an earlier standard, but it astounds us today. His yearlong expedition to British East Africa and the Congo produced 1,100 specimens for taxidermy. Because zoos had not yet perfected animal management, and few of the inhabitants prospered in captivity, naturalists of the period often collected carcasses for exhibition and for careful study by specialists in natural history museums. Roosevelt provided a wealth of knowledge in his material donations to the American Museum of Natural History in New York and the Smithsonian in the nation's capital. The realistic depiction of groups of animals within a naturalistic diorama was a precursor to landscape-immersive exhibits in modern zoological parks. Just as relevant to a modern generation, Roosevelt's words of encouragement are enlarged on the interior walls of the New York museum's rotunda:

There are no words that can tell the hidden spirit of the wilderness, that can reveal its mystery, its melancholy and its charm.

During Roosevelt's time, early in the twentieth century, the federal government began to annex land for protection and recreation. His legacy can be measured in acres. He protected 230 million acres during his presidency, including 150 national forests, 51 wildlife refuges, and 18 monuments, such as the Grand Canyon National Monument. He established the nation's fifth national park at Crater Lake, Oregon, in 1902, and he founded the National Park Service to further conservation on federal lands. Today, municipal, state, and federal governments often enter into private-public partnerships to secure and to protect wildlife habitat. Philanthropy is an essential component of any strategy to protect intact ecosystems for the future. Conservation International (CI) and the Nature Conservancy are two of the most active and committed nonprofit environmental organizations engaged in the purchase and protection of threatened and endangered habitat around the world, but many local, regional, and state land trusts are equally committed.

Today, environmental leadership is expressed by governments, corporations, nonprofits, and individuals any of which (or whom) can take the lead in a private-public partnership. Increasingly, partnerships on the environment are multigovernmental among towns and cities; county or regional governmental bodies; state and federal agencies; or legislatures. An enduring example is the partnership to protect the surrounding land and waters of the Chattahoochee River flowing through the southern states of Georgia, Alabama, and Florida.

In 1990, the river was identified as one of the ten most

endangered rivers in the United States. Because Newt Gingrich represented a key section of the Chattahoochee, its significance locally and regionally made it a high priority for a unique private-public partnership. Legislation in the House of Representatives in 1996 appropriated $25 million from National Park Service sources to enhance the Chattahoochee River National Recreation Area near Atlanta. This legislation had provisions for matching grants from the private sector. The funding expanded to a $160 million land acquisition fund consisting of private and public funding and a list of partners, including the National Park Service, the U.S. Forest Service, the Georgia Department of Natural Resources, the Atlanta Regional Commission, seven Georgia counties, and seven Georgia cities.

With assistance from the private, nonprofit Trust for Public Land, the campaign for land protection ultimately preserved 146 miles of Chattahoochee riverbank. Private-public partnerships are working throughout the nation in projects designed to protect our natural resources from the pressure of urban sprawl and the relentless growth of our cities. As we identify opportunities to purchase and set aside forested and wetland buffers around the nation's essential rivers, streams, and lakes, organized fundraising will become a top priority.

The distinction between one-time charitable giving and philanthropy should be noted. Philanthropy is organized, strategic giving to solve longstanding problems. Collaborative philanthropy, when paired with governmental or business investments is a powerful incentive for constructive dialogue on the environment. As a social cause, the environment is not yet as compelling as education or health care, but it will gain parity as environmental issues and controversies grow worldwide. Philanthropic leadership is not burdened by hardened political boundaries, so warring factions can migrate to a more flexible position of collaboration. The action orientation of philanthropy is a good way to defeat the inertia of bureaucracies.

Citizens are eagerly responding to philanthropic environmental and conservation opportunities, and overall corporate giving is on the upswing, with a total of $3.6 billion of donations from the nation's 2,600 corporate foundations in 2005, an increase of nearly 6 percent over the previous year. Corporate giving by affiliated outside foundations added another $13.8 billion to the total. Corporations gave 22 percent more than the previous year when non-foundation gifts are counted. Natural disasters, such as the Asian tsunami and Gulf Coast hurricanes of 2005, tend to increase private philanthropy. In 2005, the Bill and Melinda Gates Foundation paid out more than $1.4 billion to worthy recipients, mostly human health–related charities. Warren Buffett added $31 billion to the Gates Foundation to ensure that his money would be used effectively in an effort to enhance the world we live in. As he sat alongside Buffett, Bill Gates proclaimed "there is no reason that we shouldn't be able to cure every one of those top twenty diseases." Some of this new money may be allocated to support health-related environmental problem solving, as the Bill and Melinda Gates Foundation has made grants in 2006 and 2007 to support the Trust for Public Land, the United Nations Environmental Programme, and entrepreneurs developing innovative water technologies in countries such as Ethiopia, Myanmar, Nepal, and Zambia.

A huge commitment by billionaire entrepreneur Richard Branson will help fight the effects of global climate change. Branson will devote $3 billion over the next ten years to research and development in renewable energy initiatives. Virgin Fuels, the investment unit, expects to turn a profit by developing new and cleaner fuel technologies. His approach demonstrates that a philanthropist can direct his assets to make a difference without divesting his personal financial stake in the outcome. This may turn out to be one of the more effective ways to change human habits. Certainly, Branson has retained powerful leverage over the innovation he is seeking to facilitate. Branson has become a serious stakeholder in the business of renewable energy.

American entrepreneur Steve Case and his wife, Jean, founded the Case Foundation in 1997. Their mission "to achieve sustainable solutions to complex social problems by investing in collaboration, leadership, and entrepreneurship" is already getting results. One example of how they have successfully merged their keen business acumen and commitment to social responsibility is a new partnership between the Case Foundation and the U.S. government. This innovative collaboration is working to install 4,000 water pumps in ten African nations, bringing clean water to 10 million people. The water pumps, developed and distributed in South Africa by social entrepreneur Trevor Field, are operated by local children who turn a merry-go-round at playtime. A form of windmill technology, PlayPumps generate up to 1,400 liters of water per hour, producing a larger volume of water than a traditional hand pump with less effort. The PlayPump system exemplifies the widespread and wonderful impact of a truly creative idea.

Average Americans continue to give to causes they believe in, according to recent data collected by the inaugural Freelanthropy Charitable Giving Index. The survey examined the types of charities that were supported. Young people (18–24 year olds) tend to support health and human services causes, while slightly older Americans (25–34 year olds) rank education first on their list of priority giving. Americans aged 45 and older reported that religious organizations were their highest priority. In terms of gender differences, men preferred educational causes more than women (22.5% to 12.3%), while women ranked the environment higher, by an 18 percent to 11.6 percent margin. The consistently successful fund-raising campaigns of the World Wildlife Fund, CI, and the Wildlife Conservation Society demonstrate that environmental philanthropy is alive and well.

A large endowment gives an environmental organization an opportunity to think long term. One of the larger endowments, approaching $500 million, belongs to the Wildlife Conservation Society (WCS) of New York. The WCS endowment funds more than one hundred field conservation projects on four continents and supports the conservation platform that operates within the five zoos and aquaria that WCS operates in New York. There is no conservation-oriented nongovernmental organization that has supported field conservation longer than the WCS. The endowment is testimony to the high esteem that WCS enjoys among environmental philanthropists. The WCS also has successful partnerships with other nongovernmental organizations, which makes it possible for the organization to take on challenging environmental issues around the world. In this way, conservation is on the front lines in developing countries, doing a world of good in many places that the American government is not welcome. A future American government will find a way to support the important field endeavors of WCS, CI, Nature Conservancy, and other nongovernmental organizations that are making a difference.

The U.S. government operates endowments for the humanities and the arts. In addition, the Institute for Museum and Library Services provides grants to stimulate improvements in our nation's museums, gardens, zoos, and aquariums. Perhaps, it is time we consider a new endowment for conservation and the environment. A fund such as this could be established from both public and private contributions and broadly applied to stimulate innovations that serve our environmental priorities. Environmental priorities could be strengthened and extended with dedicated endowment funds. Another possibility would be to merge smaller foundations into a larger foundation that would grant more substantial gifts.

Dedicated organizations currently operating in the environmental domain, such as the National Fish and Wildlife Foundation, have stimulated effective private-public partnerships. Many of these partnerships have invested in conservation and the sustainable use of natural resources. Since its inception by a congressional act in 1984, the National Fish and Wildlife Foundation has awarded 7,000 grants to more than 2,600 organizations in the United States and abroad, leveraging more than $300 million in federal funds for a total of more than $1 billion in conservation funding. Similarly, several smaller, specialized funds in the Department of the Interior provide support for flagship species (e.g., the Rhino-Tiger Fund and the African Elephant Fund) that often results in private-public partnerships to protect wildlife populations in range countries where the animals live in abundance. The membership of the Association for Zoos and Aquariums has submitted large numbers of petitions from zoo visitors to request a significant increase in the funding of federal endangered species programs such as the Rhino-Tiger Fund. Although it may seem unfair to concentrate so much attention on charismatic megafauna, most conservation and zoo biologists recognize that these taxa are compelling and irresistible to donors and decision makers. If we can protect eagles, tigers, and bears, then all the smaller animals living within the same ecosystems will be protected also.

Endowments can be an important source of dedicated, long-term funding to protect endangered species and endangered places. Significant resources should be set aside to protect hallowed sites such as the Galápagos Islands, the Virunga Volcanoes in Central Africa, and Georgia's Okefenokee Swamp, to name a few. With assistance from Goldman Sachs, a global investment banking firm, WCS is protecting Karukinka, 680,000 acres in southern Patagonia, on the island of Tierra del Fuego in Chile. These and many other at-risk ecosystems will survive best with permanent, stable funding that can grow from astute investments and continuous cultivation.

Some nonprofits personalize individual animals and provide endowments or adoptions on their behalf. Donors to the nonprofit Dian Fossey Gorilla Fund International can "adopt" a wild mountain gorilla. Likewise, whales could benefit from an individual's or a company's commitment to responsible and personal stewardship, similar to state adopt-a-highway or adopt-a-school programs. We need new forms of vigilant human presence to keep wild-life and wild places safe and secure. If the University of Southern California can endow a quarterback, why can't we find a way to endow a pack of wolves at Yellowstone Park, a pair of whooping cranes in Texas, or bald eagles in Maryland?

Independent private foundations have strengthened the environmental movement in a big way since the first Earth Day in 1970. The Bullitt Foundation, for example, headed by Denis Hayes, chief organizer of the first Earth Day, made a $1 million grant to support the Earth Day Network in 2000. The Bullitt Foundation continues to help keep Earth Day a highly visible moment each year. Similarly, the Northwest Ecosystem Alliance helped to develop the Loomis Forest Fund, which raised $16.4 million to buy up logging rights to 25,000 acres of forest land in the state of Washington. Although some environmental founda-tions originated decades ago in response to serious dis-putes over land development and conservation, today's entrepreneurial foundation is more likely to engage in a win-win partnership with socially aware companies eager to earn public respect by contributing to environmental stewardship.

Established businesses have also begun to generate new private ventures to bring about rapid social change, a response, in part, to the slow pace and relatively lowered expectations of nonprofits. Investments in profit-making social programs create local business opportunities, and profits can be reinvested to help more people and groups with an entrepreneurial bent and a social conscience. Google (Google.org) has invested millions of dollars in companies and in investment funds that produce a social benefit. The Google Foundation is focused on global development, global public health, and global climate change, supporting the work of more than 2,100 nonprofits operating in sixteen countries to date. The company's founders expect the foundation to eclipse Google eventually in its positive impact on global events.

A single individual, passionate about a cause, can make a tremendous difference with carefully directed, strategic philanthropy. Bill and Melinda Gates are good examples, but a lesser-known figure is philanthropist Fred Kavli, who is funding basic research for fairly open-ended projects in nanotechnology, neuroscience, and astronomy. He has launched fourteen research centers based at Yale, Stanford, Harvard, MIT, and Caltech. Kavli requires that recipient institutions match his gifts. This requirement ensures that universities are committed to funding the Kavli science teams for the long term. In addition, he provides follow-up funding through Kavli Prizes and regular gatherings for researchers. His approach has been compared to how businesses brand products.

Kavli's management team is lean, and he doesn't enforce any particular direction for the groups he funds. Instead, he supports them in whatever direction they want to take their research, and he expects them to plot their own creative course. Kavli believes his approach opens the door for breakthroughs that will help solve the world's most intractable problems. A strategy akin to Kavli's is needed to encourage environmental research that facilitates large-scale changes in the way we use energy and our propensity to consume natural resources. Moreover, Kavli's investments in basic science such as nanotechnology will certainly lead to innovations and promising applications in environmental technology.

Philanthropy, which means literally "the love of humanity," succeeds by cultivating public awareness and motivating commitments by others. Much of the effort to advance social causes such as the environment is based on principles of "social marketing." The construct has been defined by marketing gurus Philip Kotler and Gerald Zaltman as a way to influence social behavior, not to benefit the marketer but to benefit the target audience and society. For example, Kotler and Roberto have explained that

> A social change campaign is an organized effort conducted by one group (the change agent) which attempts to persuade others (the target adopters) to accept, modify, or abandon certain ideas, attitudes, practices or behavior.

A remarkable social marketing campaign unfolded a few years ago when the African Wildlife Foundation (AWF) engaged a powerful West Coast advertising firm to produce an awareness-building series of television ads to dissuade consumers from purchasing products made from ivory. The ads depicted elephants as they were shot to death in a culling operation. The shocking images were punctuated by the slogan "only elephants should wear ivory." The AWF ads were aired on media mogul Ted Turner's TBS cable station only after Turner agreed to bypass the station's ban on depicting animal cruelty. Once Turner opened the door to this social marketing ad, other networks followed, and the production had a lengthy and successful run on television. Pro bono contributions of television or production time enable social marketers to reach a massive audience at little cost. A sustainable environmental effort in America will require a major commitment to marketing.

The creativity behind social marketing can be alarming. A recent television ad depicted water disappearing into a storm drain as a voice warned that lawn fertilizer in the spring can wind up in the Chesapeake Bay. "No crab should die like this," the announcer opines. Later, the announcer appears on screen with a small tub in hand, exclaiming "they should perish in some hot, tasty melted butter!" This promotion by the Chesapeake Bay Program, a subsidiary of the U.S. Environmental Protection Agency, promotes the Maryland region's seafood as a reason to protect the bay. One ad proclaimed: "The lunch you save may be your own." The designer of the ads explained that he intended to reach people who had not been reached in twenty years of traditional environmentalism. As creative people enter the social marketing arena, humor will be evident in the ads that break through to register awareness, change behavior, and stimulate action on behalf of the cause.

What will it cost to protect our planet from the inherent dangers of our complex geophysical environment? No expert has projected a meaningful figure, but it will be a costly venture indeed. The upside of conservation efforts is similar to Branson's investment in fuel technology: Money can be made as we clean up the earth, but not without risk. However, it will take a strong combination of individual, corporate, and government investment and philanthropy to advance a unified, global environmental strategy. We won't be motivated by profit alone. We will endeavor to cleanse the earth because it is the right thing to do.

One model for the "new philanthropy" is the Clinton Global Initiative (CGI) that aims to accelerate active philanthropy, especially when it is confronted with global emergencies such as the Asian tsunami or Hurricane Katrina. The CGI organizes a creative marketplace by gathering together experts and donors and requesting that each attendee pledge their resources, time, or leadership. Branson's pledge to fight global warming was launched with considerable media fanfare at the 2006 meeting of CGI. President Clinton's strong leadership style receives plenty of media attention for his program of high-level strategic philanthropy.

Philanthropy can play another important role in society. The philanthropic community, an alternative to both market-based and bureaucratic institutions, is uniquely capable of bringing together diverse sectors and antagonistic factions for the common good. We have consistently argued that cooperation, compromise, and civil dialogue on the environment are needed. Philanthropy has the appropriate flexibility and tempo to unite government, business, and environmental groups to overcome stalemate. Political ideologues on the left and the right will eagerly support an opportunity to help allocate private money. To make a quick, strategic impact on the environment, a diversity of politicians should be enlisted to interact with leaders in philanthropy who have resources to share. We believe that an opportunity of this magnitude will culminate in common ground and resonate with common sense.

TALKING POINTS

1. Private funding supports protection efforts through the purchase of land for environmental buffers such as wetlands and riverine forests.

2. Environmental philanthropy has reached historic levels, and donors have not reached a point of fatigue in their giving habits.

3. Keystone species, habitats, and ecosystems need to be identified for major public-private endowments that dramatically improve the protection and restoration of wildlife hotspots and corridors.

4. An organized, national program of social marketing may be necessary to boost overall giving to the environment.

5. An effective way to combat polarization and advance the common good is for philanthropic groups to invite political adversaries to consider environmental threats and a collaborative, strategic approach to funding their resolution.

8

Renewing the Natural World

Few are altogether deaf to the preaching of pine trees. Their sermons on the mountains go to our hearts; and if people in general could be got into the woods, even for once, to hear the trees speak for themselves, all difficulties in the way of forest preservation would vanish.

JOHN MUIR

JOHN MUIR, FOR MOST of his adult life, experienced nature in spiritual overtones. His intense advocacy for the Sierra landscape culminated in his role as the leader of a morally inspired preservation movement based on aesthetics. He and his followers resisted efforts to use the land in western national parks and forest reserves. Muir, the founding president of the Sierra Club, was particularly adamant about destructive activities such as grazing rights for sheepherders. He favored public ownership of forest preserves, where all cutting and economic activities were prohibited. He even favored using the armed forces to enforce environmental law. That sheep consumed and trampled forest seedlings incited Muir to label them "hoofed locusts." In hindsight, it is easy to recognize the perceived value of forest aesthetics. Beyond their economic value, forests also provide opportunities to experience the psychological effect of pristine nature. However, with the inexorable advance of the urban realm, the defense of forested lands assumes a greater urgency as we strive to sustain the highest-quality standards for our nation's air, water, and land. Many of our most populated cities are becoming barren hardscapes, devoid of green space, urban forests, or protected parks, which contributes to our growing estrangement from nature. Beyond the effects of deforestation on our psychological well-being, the loss of a shade-giving canopy and functional watersheds can compromise our physical health. Visionary environmental psychologists such as Robert Sommer have warned about the implicit connection between "hard architecture" and behavioral pathology, so we need to consider carefully ecological factors that soften our living environments.

The world's forested ecosystems are verdant masterpieces, worthy of appreciation and teeming with inspiration, but they are more than that. Our forests also contain the arteries that deliver essential nutrients. Human health, indeed the health of all living things, depends on a functional, nurturing ecosystem; water, soil, plants, and animals form a complex web of mutual codependency. By protecting the Sierra mountain ecosystem so explorers could enjoy it, Muir also reached out to protect its capacity to flower and regenerate. In the modern world, a century after Muir's time, we have witnessed world deforestation on a massive scale. From 1991 to 2000, deforestation in the Amazon increased 35 percent, an area the size of Portugal, with most of the forest losses due to conversion to pastures. The diminished forest that remains intact worldwide is the last refuge for desperate urban colonists who seek respite in its remote solitude and comfort in the cacophonous din of its splendid biodiversity.

American forests are healthier than most. America has more trees now than in 1900. Today, about 33 percent of the nation is forested; a serious comeback from a low of 20 percent at the turn of the twentieth century. One-third of the forested land is classified as "primary forest," the most biodiverse type. When the total rate of habitat conversion is calculated from 1990 to 2005, U.S. forests actually lost 0.8 percent of their woodland habitat. Given conditions in the rest of the world, stability may seem like a reasonable outcome, but we can do better at home. We can also take comfort in the fact that tree stock has increased in 22 of the world's 50 most densely forested countries since 1990. In fact there are really two world trends in forestation; a total of 69 countries where forests are increasing and 92 where forests are decreasing, according to data gathered during the fifteen-year interval from 1990 to 2005. The most troubled regions continue to be Indonesia, Nigeria, the Philippines, and Brazil. Based on research by Sedjo, Kauppi, and Ausubel, published in the *Proceedings of the National Academy of Sciences*, these trends confirm the relationship between a nation's per capita income and forest expansion. Wealthy countries are steadily increasing their investment in forests.

The October 2006 issue of *National Geographic* highlighted many of the most essential national parks around the world and advocated for their complete protection. The powerful photographs accompanying the article are inspiring. We agree that national parks should be protected to the greatest extent possible, but parks were established, in part, to be visited, and visitation inevitably affects parks on many levels. More than 3 million people visit Yosemite National Park annually. At peak times, the park can be uncomfortably crowded, even dangerous, so a sophisticated management approach is required. It may be tragic, but national parks in many countries require intense intervention, not unlike zoos or botanical gardens. The wild is less wild than it used to be, and this trend will probably continue as people seek intimate contact with nature. A frightening problem is the number of national parks worldwide under pressure from military or insurgent occupation, for example, in Congo and Myanmar. In such places, wildlife is a daily casualty of war, famine, and chaos.

The current debate about park management mirrors the Muir-Pinchot debate that arose during Teddy Roosevelt's presidency and continued for many years while Gifford Pinchot headed the National Park Service. The ultimate utilitarian, Pinchot believed that parks and reserves had to be managed. In fact, any park that experiences heavy human visitation will not be successful unless it is managed and managed well. African national parks require continuous vigilance to control littering, vandalism, and wildlife poaching. Nairobi National Park in Kenya suffers from graffiti written large on yellow-barked acacia trees, while littering has turned many wild raptors into aggressive pests, more pigeons than predators. Few national parks in America have escaped the pockmarks of graffiti artists who paint boulders and cliff sides with impunity. Our national parks and wildlife reserves require more vigorous protection. With an eye toward strengthening his environmental legacy in the 2007 federal budget, President Bush requested the largest increase for national park funding in our nation's history. We should expect nothing less than excellence from our national park system.

Stringent management practices are evident abroad in a new conservation strategy developed by the Wildlife Conservation Society (WCS) of New York. The Tigers Forever program sets unique performance standards for conservation. Biologists at WCS want to increase the number of tigers at Asian research sites by 50 percent over the next decade. The WCS's willingness to be accountable to a goal is winning support from private donors. Michael Cline, who contributed $10 million to the program, commented:

> More organizations should set goals like this. As a venture capitalist, I believe in fact-based judgments based on likelihood of success.

Because protecting prey species can save carnivorous tigers, WCS encourages incentive-based actions to cut poaching, including payments for exposing illegal hunting, bounties for captured guns or tiger traps, and bonuses for diligent action by park rangers. The emergency of looming extinction will doubtless require increasingly tough conservation action in the days ahead.

Inside the boundaries of a protected park or reserve, civilization is left far behind. Buffered by nature, we are grateful for moments spent untouched by urban pollution. In the urban realm of southern California, smog is the ubiquitous irritant. Also known as "bad ozone," smog is ozone formed near ground level composed of pollutants emitted by vehicles, power plants, industrial boilers, refineries, chemical plants, and other sources that react chemically to sunlight. Smog irritates the human respiratory system causing coughing and chest discomfort. It also burns the eyes. It can make it difficult for people to breathe normally, a problem for outdoor workers, runners, and walkers. Cities that generate high levels of ozone are obligated to disclose conditions to the public using a scale known as the Air Quality Index, or AQI. Cities such as Los Angeles advise children not to play outside on days with unusually high AQI.

Los Angeles air quality has improved in recent years because of new regulations on automobile and plant emissions, but the American Lung Association still ranks the Los Angeles area as the metropolitan region in America most polluted by year-round particles, short-term particles, and ozone. This persistent ranking contrasts with other data on Los Angeles's environmental progress. For example, the number of "exceedance" days in the Los Angeles / Long Beach area of southern California (days that ozone exceeded the acceptable levels) in a recent three-year period (1997–1999) fell to an average of 23 days from a total of 154 days in the 1980–1982 period. This is a significant decline, and new technology can take most of the credit for this change. For example, it is estimated that it would take twenty of today's new cars to generate the equivalent air pollution generated by just one mid-1960s car. Los Angeles is an interesting case for another reason: Its local government is seeking greater federal regulation than is currently mandated by federal policies. The disagreement centers on emissions generated by certain mobile sources, for example, locomotives, cargo ships, and airplanes that are not under the control of the South Coast Air Quality Management District. The struggle to determine the appropriate level and source of regulation is debated regularly in California.

Further advances in fuel technology, hybrid cars, biodiesel, and cellulosic ethanol should advance the generally positive trend in Los Angeles in the next few years. Our attention to the nation's environmental problems in the latter two decades of the twentieth century have resulted in many milestones, including the successful cleanup of the Great Lakes, the removal of lead from gasoline products, and the nationwide phaseout of chlorofluorocarbons, an indication that government and industry can act decisively when there is a demonstrated need.

Although many Californians have migrated to Oregon and Washington in search of cleaner air, escaping smog is not that easy. Other American cities, from Seattle to Atlanta, and many other locations throughout the country, have experienced similar problems with ground-level ozone. The American Lung Association estimates that 159 million Americans live in counties with unhealthy levels of either ozone or short-term levels or year-round levels of particle pollution. We have been fighting smog in American cities for thirty-four years and have succeeded in lowering average ozone levels approximately 24 percent between 1980 and 2000. However, efforts to relax the federal standards will not help to defeat the anomaly. Even if government and environmental activists have overestimated the problem, we still have a long way to go to certify that all Americans are breathing easily.

Our Contract with the Earth requires a national commitment to clean air, land, and water, so we counsel unwavering vigilance on this issue. The increasing public demand for improvements generates an abundance of creative, pragmatic solutions. For example, river restoration in Los Angeles helps to control toxic dust storms resulting from unsound water policies that long ago dried up Owens Lake, an area more than three times the size of Manhattan. The lake's basin is the largest single source of particulate matter pollution in the nation, and a distinct example of how water ecology can affect the quality of our air.

The value of urban forests and streetscapes of living plants is measured in part by their ability to sequester many pollutants, including nitrogen dioxide, sulfur dioxide, carbon monoxide, ozone, and particulate matter ten microns or less in size. This is a significant contribution to air quality. For example, experts in the U.S. Forest Service estimate that 19 million pounds of pollutants in Atlanta and more than 1 million pounds in Denver are removed annually from the air by resident trees. The value of this filtering process each year has been calculated at more than $50 million. The increasing loss of trees in American cities and towns reduces the effectiveness of this natural barrier to air pollution. The steady increase of tree plantations in rural sites won't help America's cities escape ozone. Heavily populated Asian cities such as Beijing and Taipei, where finding a single living tree is a challenge, have some of the worst air pollution in the world. Tragically, deforestation has many manifestations that work against world health. Cleaner cars and more trees are a fine combination for fighting ground-level ozone.

Worldwide, but especially in America, nonprofit environmental groups, service clubs, churches and synagogues, scouts, and local governments have successfully waged tree-planting campaigns. A visible tree-planting campaign is one of the most motivating and inspirational activities that Americans can do for their communities. Any shape, size, or type of tree can help to absorb carbon dioxide. The National Arbor Day Foundation provides leadership to encourage tree planting throughout the nation. This kind of entrepreneurial intervention to revitalize and reshape our urban landscape can be encouraged by prizes and grants awarded to community service clubs, scouting organizations, and schools. Philanthropy influences the rate and breadth of participation by the targeted provision of recognition, incentives, and rewards. Impoverished communities in both urban and rural settings should be our first priority.

Wangari Maathai grew up on a family farm in Kenya and studied in the United States. When she returned to her country, she recognized that the spiraling poverty among her fellow citizens was linked to the rapidly degrading environment. To demonstrate her concern, she started the Green Belt Movement, which resulted in an organized effort to plant more than 30 million trees in Africa, with phase one firmly established in Kenya. The effort aims to stabilize the soil, control erosion, and provide jobs for people. She won the Nobel Peace Prize in 2004. Tree-planting is an extraordinary act of environmental heroism. The recognition that Wangari Maathai has received demonstrates how much forestation is valued throughout the world. Tree planting is valued in America as well, as New York Mayor Bloomberg confirmed with his pledge to plant more than 1 million new trees in Manhattan.

Trees are not the only contributors to clean air; there is new technology on the way to help. The Environmental Protection Agency predicts that during the next two decades average motor-vehicle per-mile emissions will drop by 85 percent because of cleaner vehicles on the road. Cleaner industrial and homestead technology will enable Americans to enjoy their suburban lifestyle as they modify their daily habits. Development is often tied to strict clean air standards, so the stakes are not trivial. Our nation's economy cannot afford to be halted by faulty calculations. Entrepreneurial environmental breakthroughs may relieve some of the heavy-handedness of government if air pollution continues to improve. However, Americans will have to export their environmental technology in a big way to offset current world trends in air pollution. In Mexico City, for example, old cars banned in Los Angeles, numbering more than 3 million, are spewing massive quantities of pollutants with no signs of legal restraint. At 179 milligrams per cubic meter of suspended particulates, Mexico City's air pollution is twice the maximum levels recommended by the World Health Organization.

Bad air can migrate, so ozone, like acid rain, can spread from one country to plague its neighbors. One remarkable development in Mexico is the Sistema Informacion Ambiental, a nongovernmental organization that will make Mexico the first nation to provide air quality and ultraviolet data online for urban populations throughout the country. They already provide this information for Mexico City. Its user-friendly qualities won acclaim for its visionary chief executive, Luis Roberto Acosta. Growing public concern about the environment resulted in the most stringent air pollution control measures of any city in the developing world. We can hope for further environmental improvements in Mexico.

America's streams, rivers, lakes, and oceans that border our nation are also threatened by dangerous pollutants that degrade the quality of our water. President Bush's positive step to establish the world's largest marine reserve in the Northwestern Hawaiian Islands is undermined by a vortex of plastic trash, endlessly circulating within the island ecosystem by prevailing wave action. Turtles, albatrosses, and other marine species have been entangled in floating plastic. Hazardous plastic is generated on land and deposited in the open sea where it remains uncollected for years.

From baseline observations documented by environmental groups patroling the oceans, further action will be needed to keep this marine reserve pristine and protected. This is also a direct effect of our propensity to "dump" our waste instead of recycling and reusing it—a further signal that we need a comprehensive environmental strategy. Vigilant recycling efforts in every community will considerably reduce the amount of waste that ends up in the seas. The cruise industry's embrace of recycling has contributed significantly to the cleanup of our oceans. Environmental "prompts," or print and electronic media campaigns, will help publicize recycling. We know that recycling works, but we have to sustain the effort by building enthusiasm and morale—subdivision by subdivision and business by business—another important role for our local, regional, and state government leaders.

An example of a successful environmental prompt is the Don't Mess with Texas antilittering campaign. This public relations effort, initiated in 1986 to reduce the cost of roadside litter pickups by government, was honored

as American's favorite slogan in a national contest for
the Advertising Week Walk of Fame. Five years into the
campaign, Texas realized a 72 percent decline in roadside
litter. The slogan was so compelling that 73 percent of
Texans could identify it within a few months after the campaign was launched. Clever marketing will be necessary to
encourage best environmental practices by American citizens preoccupied by a busy marketplace of ideas and information. Fortunately, American business has experience
with persuasive market techniques. We must be willing to
deploy our best and brightest marketing geniuses to drive
citizen participation and responsible, mainstream environmental activism. Fellowships to provide graduate-level
training in social marketing are a good place to start.

New technologies are advancing to the marketplace,
but consumers can be wary of change. If they unequivo-
cally demand alternatives to gasoline, for example, the
marketplace will be activated, but there is plenty of resistance to overcome. Recently, the president of Shell Oil,
John Hoffmeister, expressed his views on world demand
for energy and business opportunities ahead. Shell Oil's
position, as Hoffmeister explained it, is that America will
always need foreign oil even as it aggressively develops
alternatives such as solar, wind, ethanol, gasification, and
hydrogen sources of power. This view contradicts those
who have strenuously argued that our national security
requires an end to foreign oil dependence, especially in the
Middle East. It is difficult, if not impossible, to have an

intelligent conversation about energy if we cannot agree that independence from foreign oil or, in the very long run, independence from fossil fuel, is achievable. In the best of circumstances, it will take time to end our dependence, but many experts believe it is time to commit to this worthy goal. Our national security and our nation's economy depends on supply and demand shifts that are orderly, predictable, and carefully managed. Chaos is the enemy of national security.

By reducing expectations, it is easier to enable fossil fuel to retain its grip on the American consumer, and it becomes more difficult for a competitive industry to gain a foothold, even if it is an industry spawned by Big Oil. When Shell Oil announced its plan to invest in wind and solar power, skeptics wondered whether Shell Oil's commitment is long or short term, a tactical diversion from the core-business objective to prolong and extend the use of fossil fuels.

Our nation will need tax incentives to continue research on alternative fuels. By speeding up research and development, a new breakthrough in automotive technology based on hydrogen, for example, could potentially end our reliance on gasoline. If we commit to the challenge, an energy revolution is not far off. Right now, hybrid cars look like the front runners for market share in alternative fuel. Plug-in hybrids are particularly attractive because they permit the user to recharge batteries at home where solar or wind power can be deployed or use the grid during off-peak times to lower energy costs. A hybrid car with a sixty-mile range might not use any gasoline during short-distance commuting to and from the workplace.

National security is driving public concern about our nation's commitment to fossil fuels. A prominent group of corporate CEOs (FedEx, UPS, Dow Chemical) and well-known, retired generals have urged the White House to adopt new energy policies that reduce our dependence on foreign oil. In a letter addressed to the president, Congress, and the American people, representatives of the Energy Security Leadership Council argued that American dependence on foreign oil makes our country vulnerable to terrorist attacks. For example, America's transport system functions nearly 100 percent on petroleum, whereas 90 percent of the world's oil supply is controlled by foreign governments. Formation of this coalition of business and retired military officers was designed to break the current partisan stalemate in energy policy. They advocated improvements in fuel efficiency, the development of alternative fuel sources, and increased exploration and production of fossil fuels. Recognition that hostile states and terrorist groups intend to use oil as a strategic weapon against American interests elevates oil dependency to an urgent priority. As the coalition spokespeople concluded:

America's oil dependence threatens the prosperity
and safety of the nation. Continued policy paralysis
is unacceptable precisely because we can take action

to improve our energy security. Many challenges lie ahead, but we have no doubt that the efforts of the American people will meet with success.

Our country will need the equivalent of a national commitment for dramatic growth in alternative energy, enough growth to sustain investments and expand market share for these new, cleaner, homegrown industries. Established oil companies will find a way to enter and succeed in these new businesses, but they must also consider the possibility that the oil business, as we know it, may not survive the competition. That outcome, if it happens, is a long way down the road. Americans should root for the survival of these industrial giants, whose products may be reincarnated in an earth-friendly, renewable form.

Shell Oil Company is on record for acknowledging the significance of global climate change. David Hone, Shell's Group Climate Change Adviser, stated that climate change requires a fundamental strategic shift in responding to future energy needs. As he further concluded:

An expansion of today's energy infrastructure to meet future demand without also encompassing a change in its carbon profile will not be sustainable.

Hone also noted that, despite a pattern of active investment in alternative energy sources, the pace of investment is too slow, another example of industry's readiness to move forward on environmental issues even when economic conditions are not favorable. We continue to be optimistic that business has much to gain from innovation that protects our environment.

Many companies, including Shell, British Petroleum, and Florida Power & Light, are increasing their investments in solar and wind power. Wind-driven energy sources are gaining support in the United States. In Texas, for example, then-Gov. George W. Bush approved the Texas Renewable Portfolio Mandate, which stipulated that state power companies should produce 2,000 megawatts of electricity from renewable sources by 2009. Many new companies competed to build wind turbines so that the goal was reached four years earlier than mandated. Wind power in Texas is now competitive with clean coal, nuclear energy, and natural gas.

Until recently, solar power was viewed as too expensive, but photovoltaic technology is rapidly advancing as a cost-effective, efficient energy source. Within a decade or less, solar energy may be able to compete with other sources of energy. A serious effort to fund solar research and development, on the scale of the World War II—era Manhattan Project, would doubtless yield huge progress in the production and marketing of solar energy throughout the nation. Our federal government should take the lead on this vital issue, an effort that may require strong incentives to encourage enterprise and drive the formation of private-public economic partnerships.

Solar's potential on a small scale is apparent as green-oriented small businesses and nonprofits (especially museums, gardens, and zoos) attempt to meet LEED (Leadership in Energy and Environmental Design) certification for their newest buildings. Major corporations such as American Honda and Steelcase Corporation have also achieved LEED standards in their newest buildings. In fact, according to green design consultant Greg Kats, about 500 million square feet of green buildings are under design, development, and construction in America.

Honda's Northwest Regional Facilities in Gresham, Oregon, use 120 skylights designed to keep light levels up and costs down. During sunny days, artificial lighting in the building is unnecessary. The building was designed to use less than 40 percent of the normal energy of a building of this type based on the Oregon Energy Code. Honda also committed to using low-energy input materials, recycled products, and renewable sources in the building process. Buildings are a significant contributor to America's energy demand. Experts estimate that commercial and residential buildings in America are consuming 65 percent of our electricity resources, 12 percent of potable water, and 40 percent of all raw materials in construction. Clearly, the trend to greener building methods has much to offer as we strive for healthier neighborhoods, towns, and cities.

An important property of LEED certification is the requirement that LEED buildings are used as education centers to influence the adoption of green building technology. LEED certification may provide local and global education to sustain green research and development worldwide.

Solar power is a simple but powerful way to meet exemplary standards for energy efficiency and waste reduction. If America can lead the way in the development of competitive solar technology, the rest of the world will follow. The effects of widespread solar use will be particularly important in China and India where air, water, and soil pollution is out of control because of rapid industrialization and urbanization. Forty-five cities in China have populations that exceed 1 million people, while India is expected to have seventy cities with similar-sized populations by 2025. China's demand for oil has doubled since 2000, accounting for 40 percent of the increase in demand worldwide. Similarly, the demand for energy in India is increasing annually by a factor of 30 percent.

Rapidly developing countries are faced with a choice between the traditional and inherently toxic road to development or a new approach that uses primarily renewable energy sources. The renewable alternative is nothing less than a reinvention of the western development pathway. China and India are already leaders in the development of wind, solar, and biogas power sources. For example, China now meets 10 percent of its total residential hot-water demand through the use of solar hot-water heating systems, representing 75 percent of the world's capacity. It is essential that this green trend win out in a struggle to industrialize both countries. Regressive, dirty technologies currently dominate.

With help from visionary American architect William McDonough, China is developing six model cities. McDonough and his Chinese collaborators are building low buildings against earth berms with roof gardens. Streets are oriented at a fifteen-degree angle to inhibit chilly winter wind and circulate urban air. Each building is designed to capture the maximum amount of solar energy. The Chinese government has already commissioned the world's first fully sustainable eco-city in Dongtan. The city is planned to be entirely self-sufficient in energy, water, and food when it is completed in 2040. It is expected to reach this level of energy efficiency by innovations in urban water harvesting and purification, community waste recycling, wastewater-based biogas facilities, and co-generation power plants. China is also building the largest wind power plant in Asia in Erenhot City, in China's Mongolia Autonomous Region, providing clean energy sources for Beijing's 2008 Olympic Games. The project is supported by $1.2 billion of private investment by Avalon Power Corporation of Canada. Boosted by 300 days of strong wind in this port city, the project is expected to generate over 1 million kilowatts of electricity annually.

India is making similarly dynamic progress in environmental development. The nation built the first platinum-rated LEED building outside the United States in 2004. Only two other buildings at this level exist in the world. At least ten more LEED-registered buildings are planned for urban sites in the country. Of course, India and China have engaged environmental innovation by vastly different methods. India relies on private investment within a democratic governmental process, whereas China's centralized communist government invests vast sums of state money and forcibly directs the participation and translocation of its citizens according to its definition of the common good. Capitalism operates in both nations, but government is a far bigger player in China. Clearly, people need to participate in decisions to clean up the environment, and they must be educated to understand their entrepreneurial role in the emerging green revolution. China may discover that green technologies actually increase demand for democratic systems of decision making. It is legitimate to question whether a country's environmental programs can be truly sustainable without a foundation of democratic institutions, values, and principles. A sign of change in China is the recent decision to protect private property rights, a milestone toward the emergence of a true market economy.

To continue the momentum of this emerging environmental century, we must find common cause on the tough issues that have polarized political infighters on the right and the left. On climate change, for example, conservatives and many corporate leaders fear that liberals will go too far or too fast to protect ecosystems under siege by perceived irreversible forces. However, liberal politicians and many academics don't trust conservatives or industry to control or reduce carbon emissions that contribute to the warming trend most climate experts have documented. The impasse created by a lack of trust and a resistance to compromise keeps conservation issues on the back burner. Still, in a command and control environment with mandated caps on carbon dioxide emissions, Europe lags behind the free-market American achievements, with Europe emitting carbon dioxide at a growth rate three times that of the United States, according to data gathered from 2000 to 2004.

National media are tracking some hopeful trends in our nation. *Newsweek* published two environmental editorials in a single issue, one by Al Gore and the other by Richard Branson. Gore's piece extols the virtue of small, diversified, and renewable sources of energy, for example, windmills, solar photovoltaics, and ethanol and biodiesel fuels. He envisions an "energy electranet" that feeds a smart electric power grid that gives unprecedented autonomy to users. As Gore points out, market forces are acting on these opportunities as companies such as Wal-Mart, BP, and General Electric get into the alternative energy game.

However, Gore also envisions the need for real leadership to hasten the world's entry into a future that is no longer dependent on petroleum for power. In our view, strong leadership will work best when it raises awareness and participation and encourages the marketplace to generate ideas and technologies that solve environmental problems. Strong leaders must avoid an overly punitive social climate so that governmental incentives and direction coexist with successful entrepreneurial business interests. Entrepreneurial environmentalists will produce the new energy paradigm for America.

Gregg Easterbrook's optimistic book, published in 1996, provides considerable support for the vision of entrepreneurial environmentalism. As Easterbrook argued, "the Western world today is on the verge of the greatest ecological renewal that humankind has known." A few of the premises that he advanced at the outset of his important book are especially noteworthy:

1. In the western world, pollution will end within our lifetimes, with society almost painlessly adopting a zero-emissions philosophy.
2. Most feared environmental catastrophes, such as runaway global warming, are most certainly to be avoided.
3. Far from becoming a new source of global discord, environmentalism, which binds nations to a com-

mon concern, will be the best thing that's ever happened to international relations.

4. Humankind, even a growing human population of many billions, can take a constructive place in the natural order.

Easterbrook labeled the many other propositions he offered in his book "ecorealism." As the next wave of environmental discourse, the author listed its core principles as follows:

1. Logic, not sentiment, is the best tool for safeguarding nature; 2. Accurate understanding of the actual state of the environment serves the earth better than expressions of panic; 3. To form a constructive alliance with nature, men and women must learn to think like nature.

The future of entrepreneurial approaches to the environment is fascinating if you are willing to challenge your imagination. Stewart Brand, founder of the *Whole Earth Catalog* and a vintage environmentalist, envisions a new frontier of technology, including the widespread embrace of genetic engineering as a way to grow crops on less land with less pesticide, new microbes that protect ecosystems against invasive species, and bioengineered fuels. An exponent of nuclear energy, he warns us that apocalyptic thinking is a greater danger to humanity.

As 2006 ended, *Time* named "You" as its "Person of the Year," opting to recognize the explosive growth and influence of user-generated Internet content such as blogs, video file-sharing sites such as YouTube, and social networks like MySpace. As *Time*'s editors proclaimed:

> For seizing the reins of the global media, for founding and framing the new digital democracy, for working for nothing and beating the pros at their own game, *Time*'s Person of the Year is you.

Time's acknowledgment indicates how entrepreneurial behavior can quickly alter human events and circumstances. The power and potential of individual acts of enterprise is a major source for our optimism about the environment. The world can be changed faster by the spread of brilliant ideas than by any plodding bureaucracy, and we gladly put our faith in such intellectual and social processes.

TALKING POINTS

1. Deforestation may be the most serious environmental challenge in our lifetime, although wealthier and wiser nations have instituted successful reforms to encourage fast-track reforestation.
2. Big Oil's investments in solar, wind, and biofuels suggests that corporations are getting serious about alternative forms of energy. If government can continue

to support tax incentives, new technologies may succeed in capturing significant market share and turn the corner toward a cleaner future.

3. Some observers believe that nations united by the cohesive demands of environmentalism are more likely to find common cause on many other international issues and thereby avert serious conflict.

9
Leading the Way to a Better World

*The facts of nature cannot in the long run be
violated. Penetrating and seeping through everything
like water, they will undermine any system that fails
to take account of them, and sooner or later, they will
bring about its downfall. But an authority wise
enough in its statesmanship to give sufficient
free play to nature—of which spirit is a part—
need fear no premature decline.*

C. G. JUNG

BOLD LEADERSHIP WILL BE a critical factor as we attempt to build a better world, but effective leadership requires knowledge, patience, and perspective. At all levels of government, in our businesses and in our homes, we all have to filter information made even more difficult by ideas cascading from the Internet. Where do we go to find out what works before we invest our hard-earned money? Even well-established consumer services are not sophisticated enough to guide us, but the marketplace may yet yield sufficient guideposts for a twenty-first century of environmental innovations. More than ever, America must develop and nurture wise leaders and must find a way to induce the participation of our best and brightest citizens—talent that is averse to the rough-and-tumble of politics. The culture and landscape of politics must adapt to these harsh realities. Both intellect and civility must be valued and demanded for environmental politics to succeed because degradation is cumulative and we cannot renew the natural world until we stop fighting long enough to agree on the proper remedy. If we strum or spar while our communities smolder, the environmental century languishes in stalemate. This can no longer be tolerated. Leaders must be shaped with attributes that fit the awesome challenges we face. We need leaders who can unite us, not divide us. More than any other issue, the environment requires consensus, or we risk a culture in which reactive forces neutralize every advance and remedy.

In his best-selling book, *Good to Great*, Jim Collins observed that effective executive leaders channel their self-interest into the larger goal of building a great company. Their ambition is mostly for the good of the institution rather than themselves. These leaders were found to be a paradoxical mix of humility and will. They are tenacious and diligent, modest and self-effacing, and understated. Collins further observed that companies tend to look for charismatic leaders—more show horse than workhorse—but charismatic leaders don't produce the best results. By this model, we should be looking for political leaders of substance. Clearly, a sustainable environmental culture will require staying power. Today's leaders must be tenacious advocates for the natural world, driven by results and guided by evidence. Where will we find such leaders? According to Collins, they are all around us, but they don't stand out. Look for a well-run company, a high-achieving organization or government entity, he advises, and you are likely to find a superior leader, who shuns attention but gets results. By recruiting such leaders, in our companies and our communities, we create a demand for focusing on the bigger picture of humanity. Some critics have asserted that the world is in bad shape, but even if it were simply in "good" shape, who would object if we committed to taking the earth from good to great? We should renew the earth to a high standard in every way we can. Individually and collectively, leadership will contribute mightily to this hopeful transformation.

One-dimensional political leaders are not easily elected, but the environment is an issue that today's candidates cannot ignore. "Quality of life" requires clean water and air and a healthy ecosystem of plants and wildlife. There is a spiritual and a patriotic aspect to environmental protection with a broad base of support throughout America. The environment has also become a hot button for Americans concerned about our national security. Americans strongly agree that we must protect our natural assets and national treasures for future generations.

We wouldn't be surprised to see a future act of terror enacted against broader, environmental targets. The Taliban exploded massive religious totems in Afghanistan, and Saddam Hussein destroyed wetland habitats to punish his political enemies in Iraq. Nuclear terror will have profound environmental effects, just as the aftermath of 9/11 unleashed dangerous particulate matter and chemical residue into the streets of New York. It will take decades to clean up from war and preparation for war. The cleanup is especially challenging in nations that lived in secrecy during the cold war. A recent study published in the *Journal of Peace Research* demonstrated that democracies exhibit stronger international environmental commitments than nondemocratic nations. The author, Eric Neumayer, concluded that the spread of democracy around the world would lead to an enhanced commitment to the environment. Open societies are more receptive to environmental and social reform.

Our nation will need to construct a leadership platform that unites environmental groups, political parties, educators, scientists, and business leaders. Too often these groups work at cross-purposes. Environmentalists have been known to clash in ways that are counterproductive, but harsh realities are uniting factions whose antipathy is rooted in history, as the environmental ethicist Bryan Norton observed:

> The environmentalists' dilemma . . . manifests itself in a number of ways . . . in factionalism and distrust of those perceived to have joined the other camp. Other environmentalists remain uncommitted and uneasily embody both factions as internal personae. The resulting theoretical schizophrenia can paralyze us with inarticulation and humble us.

Our political leaders must bridge the gaps and bring disparate groups together for discussion, debate, and environmental action. We will need to organize a series of high-level global conferences that focus on specific problems, but we must be inclusive as we engage a diversity of stakeholders. This approach will be particularly important as we consider the prospects and economic consequences of a broad environmental coalition. Our enthusiasm for an entrepreneurial environmental culture acknowledges that environmental commitments will be costly, and long-term ecological interests will sometimes win out over short-term economic gains. This is why we advocate long-term strategic planning on the environment. Well-run companies are already benefiting from green strategic thinking. Effective nongovernmental environmental groups, such as Conservation International and the Nature Conservancy, have learned to form partnerships to study environmental challenges; they explore every opportunity to bring business, governments, and local people to the conference table. We must also be careful to organize conferences that include a diversity of views on a subject. Without open and civil debate, these conferences are headed for trouble.

In their excellent book, *Green to Gold*, Daniel Esty and Andrew Winston identified many ways that leading-edge companies go beyond the basics of mere compliance to adapt environmental perspectives into all aspects of their company operations. These stellar companies, mindful of the increasing level of transparency provided by Internet access to the business world offer proactive pathways to an environmental business culture; for example, they have

- Designed innovative products to help customers solve environmental problems.
- Promoted better environmental stewardship by suppliers.
- Tracked the performance of company environmental initiatives.
- Partnered with other stakeholders to develop new environmental innovations.

- Set ambitious ecological goals, incentives, training, and tools to engage all employees in the company's environmental vision.

They do this because even smart companies can be surprised by environmental events, and surprises can cost a company serious financial loss. In addition, genuine benefits can result from taking a fresh look at your business whenever you are worried about unforeseen threats. As Espy and Winston asserted, "smart companies seize competitive advantage through strategic management of environmental challenges." If this is true of business, and we believe it is, it must also be true for governments, communities, and households. A strategic approach to the environment will help every stakeholder, from individuals to large populations and interest groups.

As we form judgments about political candidates, we ought to consider the qualifications of their allies and their entourage. We rarely get a glimpse of a presidential candidate's cabinet before the election, but we can take a close look at those who are working in their campaigns or appearing on television to support them. Successful companies employ the brightest people they can find. This is how they prepare to win in a busy marketplace of ideas. Governments based entirely on ideology will be too narrow to address the complicated challenges of the future; so we must select our leaders based, in part, on their willingness to reach out to a broader constituency. In essence, we regard the environment as a bipartisan issue requiring constant communication, cooperation, and collaboration among diverse constituencies. We should be willing to consider and evaluate any good idea, regardless of its origin. Political campaigns must be consistent and coherent to succeed, but creativity in politics, business, and science depends upon a diversity of ideas to create lasting solutions.

A good example of presidential leadership supported by gifted advisors is the presidency of Theodore Roosevelt. Roosevelt assigned Gifford Pinchot the task of leading the U.S. Forest Service, and he proved to be a crafty, able administrator. As Roosevelt's chief environmental advisor, Pinchot advocated conservation of the nation's forests through careful planning, systematic harvesting, and renewal. He was a visionary and a scholar, founding the Yale University School of Forestry in 1900. The connection between conservation and natural resources was literally invented by Pinchot, who was distinctly utilitarian in his views. He believed that forestry managed wisely should yield economic benefits, whereas romantic preaching was doomed to failure. He was, in essence, a manager at heart, and scientific management was Pinchot's chief conservation tool.

Pinchot's ideas represent one end of a continuum still prevalent in the environmental movement today. Some writers see this distinction as simply the dichotomy between "conservation" and "preservation." Georgia Tech professor Bryan Norton defined the former in the following way: "To conserve a resource or the productive potential of a source-generating system is to use it wisely, with the goal of maintaining its future availability or productivity." "To preserve it," he countered, "is to protect an ecosystem or a species, to the extent possible, from the disruptions attendant upon it from human use." Intense conflict has been generated from these distinctions as they are practiced today, but there is plenty of opportunity to find common ground.

Bold environmental decisions are often driven by determined individuals stepping forward to correct a deficiency or to right a wrong. One chief executive actively pursuing environmental policies can make a profound difference in a short time. Costa Rica's President Oscar Arias won the Nobel Prize for his efforts to demilitarize Central America, but he also exerted global environmental leadership and demonstrated by his country's example how people can pursue practices that are good for business and good for the environment. Costa Rica has the most successful ecotourism in the region, which brings much needed income to its citizens. As we have noted, the country has also converted many traditional industries to nontraditional enterprise with the help of conservation organizations such as Conservation International and the Nature Conservancy.

Costa Rica is a shining example of entrepreneurial environmentalism. President Arias continues to collaborate with environmental groups and international business interests to enlarge his vision. Under the terms of the Clinton Global Initiative, Costa Rica recently accepted an initial grant of $2 million to lead a coalition of nations seeking to protect their rain forests. Since the Kyoto Protocols are not comprehensive in protecting rain forests, Costa Rica and other members of the coalition will provide economic incentives for countries that refrain from cutting them down.

Market-based conservation is sometimes characterized as "ecosystem services" embodied in the example of the famed Catskill/Delaware Watershed. The watershed is widely regarded as a cheaper and superior filter for New York City water than any man-made filtration plant yet envisioned by technicians. Clearly, this natural alternative is a win-win for the citizens of New York and the nearby communities. However, Douglas McCauley of Stanford University has observed that nature is not always benign or benevolent, nor does it behave for the benefit of any single species. As he noted, trees take water from watersheds, forests may contribute to warming, and predatory animals sometimes kill livestock and the occasional human being. Furthermore, the value of nature is cyclical, and environmentally friendly businesses do not always turn a profit; some may never be profitable and force abandonment of an essentially good, green program. According to McCauley,

To make ecosystem services the foundation of our conservation strategies is to imply—intentionally or otherwise—that nature is only worth conserving when it is, or can be made profitable. The risk in advocating this position is that we might be taken at our word. Then, if there is a "devaluation" of nature . . . what are we to tell local stewards who have invested in our ideology, and how can we protect nature from liquidation?

We also have to be careful not to underestimate human ingenuity, as natural alternatives to technology may be a liability over time. If the only reason to adopt natural systems is economic, changing business conditions may eventually favor technology over nature. In addition, sometimes business interests overwhelm nature, as when alien wildlife is introduced into a habitat. For example, the fur-bearing rodent *Myocastor coypus* that escaped into the Kenyan ecosystem of Lake Naivasha and the population of Nile perch purposely introduced as a game fish into Lake Victoria produced troublesome ecological consequences for Kenya as the aliens outcompeted indigenous fauna and depleted vulnerable flora.

Our nation's current leaders surely understand that nature has intrinsic value that is impossible to quantify in economic terms. Nevertheless, according to conservation biologist Stuart Pimm, a group of collaborating ecologists and economists once estimated the value of the world's ecosystems and derived a number that was bigger than the global sum of all gross national products, amounting to tens of trillions of dollars. By this paradigm, ecosystem services were defined as the flow of materials, energy, and information from the biosphere that supports human life, including:

1. Regulation of the atmosphere and climate.
2. Purification and retention of fresh water.
3. Formation and enrichment of soils.
4. Nutrient cycling.
5. Detoxification and recirculation of waste.
6. Pollination.
7. The production of lumber, fodder, and biomass fuel.

Efforts to calculate the value of nature have produced some fascinating statistics. For example, in the Charles River Basin in Massachusetts, 8,500 acres of wetlands were preserved as a natural valley storage area to control flooding. The cost to acquire the wetlands was $10 million compared with an alternative proposal to construct dams and levees at an estimated cost of $100 million. Other experts have calculated the cost of replacing wetlands at $300 per acre foot, a high price by any standard. The economic value of nature is certainly debatable and undoubtedly gigantic, but nature conservation is ultimately a moral issue, and morality has proved to be a powerful motivator in American life. McCauley's commentary rightfully identifies several moral issues that culminated in sustainable legislation. For example, the commercial ban on whaling, the establishment of America's vast national parks and reserves, and the ban on the importation of ivory and ivory products. Voters were willing to set aside vast tracts of land to protect previously exploited species because they placed greater value on their survival. The record shows that Americans will elect public officials who support the moral foundation of conservation, habitat protection, and biodiversity.

A species can have direct economic impact if it suddenly disappears from the ecosystem. An example is the recent mysterious die-off of honeybees, insects that play an important role in agriculture. A Cornell University study determined that honeybees annually pollinate more than $14 billion worth of seeds and crops in the United States. The distribution of working bees has led to the industry of beekeeping, so the loss of bees not only affects the growers; it also affects the people who manage the working honeybee population.

It is especially gratifying to acknowledge that the case for conservation is made so easily on the basis of economic, social, and moral consequences, giving leaders from every political camp the opportunity to embrace the issue. Environmental action depends on a measure of consensus in which career politicians may disagree on the means to an end, but they have reached common ground in their commitment to the cause. Whether government or business takes the lead on environmental issues is not as important as their willingness to form an effective partnership. Our civic leaders must bring all interested parties to the table of conservation. Once seated, we face the future united by our belief in American know-how and our willingness to take on one of the toughest assignments in human history. Our resolve will be tested.

Sustained environmental protection and new, safe, and clean environmental technologies will require commitments that cross hardened political boundaries. We face funding shortfalls in energy research, which has plunged to nearly half the levels established a decade earlier. Because venture capitalists tend to fund ideas that are nearly ready for the marketplace, the type of arduous research that produces real breakthroughs can only be funded by astute governments. America used to be that kind of government, but our commitments have wavered in recent years so government incentives for energy research will be issues in future political campaigns.

We understand the need for strong, effective leadership in the implementation of our Contract with the Earth. America's next president, if not elected on a history of environmental commitment, will surely be elected on a platform of environmental promises. The time is right for environmental leadership by individuals and by the American nation, and we must lead by the example of our enterprise. By helping to restore the world's heritage of functional ecosystems, we will also hone the necessary technologies and programs that can be applied to continue its protection.

A diversity of leadership is required to engage our citizens in the environmental century's priority industry: tinkering with humanity's impact to restore and renew ecosystems. We have time to get it right if we get to work today. Democrats, Independents, Greens, Libertarians, and Republicans must find common ground on the environment. It is quite possible that the environment may become the world's most inclusive political issue.

Mobilizing other countries to join us will not be as easy as it may appear. Many of the nations that signed and ratified the Kyoto Protocol are lagging behind on their commitments. Canada's environment commissioner Johanne Gelinas, said in 2006 that her country has "done too little and acted too slowly" in addressing climate change. In particular, Canada has no adaptation plan to deal with drought, melting permafrost, rising sea levels, or storm surges, situations that require planning and preparation of population centers at risk. If Canada, a wealthy, literate, and democratic society, is not yet prepared, we can expect that global preparation for adaptation and mitigation is going to be a troublesome assignment. To be a global leader, America will have to be proactive and persuasive on a massive scale.

The entrepreneurial approach to climate change was recently addressed in a report compiled by Republicans for Environmental Protection:

America is ready to meet the challenges posed by global warming. America has the best scientists. America's businesses lead the world in developing and marketing innovative technologies that transform lives. All that remains is leadership that will channel the unrivaled power and creativity of markets toward developing the solutions we need soon to protect our atmosphere, strengthen American's economy, and bring clean prosperity to the world's developing nations.

Future environment-friendly policies must not be abandoned before they can prove themselves because a continuous commitment to innovation is essential. We will have to be patient as we invent, test, and try out new industrial methods and adopt new patterns of consumerism and consumption. Changes in political leadership cannot be allowed to disrupt strategic programs if they are working. Fortunately, there is plenty of evidence to suggest that Americans are already reaching consensus on the need for fresh environmental thinking, new metrics, and tools of evaluation, and the business and social opportunities represented by cleaner technologies and better personal practices.

We are entering a period of great change and opportunity, but our transition to a functionally fitter planet need not be marked by political turmoil. Our confidence in the civility, character, and creativity of the American people is fuel for our unwavering commitment.

TALKING POINTS

1. Executive leadership is a critical variable in the race to renew the earth. The nature of contemporary environmental challenges requires leaders who can comprehend complexity, prioritize options, and persuade the electorate to follow their lead.

2. Bold leadership produced the Endangered Species Act in 1973, perhaps the most effective piece of environmental legislation in our country's history. The act has been, by any measure, a very successful guardian of wildlife and habitat and any attempt to weaken it should be resisted.

3. The U.S. government has called attention to the problem of melting polar ice and its potential effect on polar bear habitat. Strong leadership, applied at the right moment, is often the difference between survival and extinction for wildlife.

10
From Many, One

One touch of nature makes the whole world kin.

WILLIAM SHAKESPEARE

OUR *CONTRACT WITH THE EARTH* is intended to inspire, but it may not inspire you to *feel* differently. Rather, we hope, it will inspire you to *see* differently. It examines the landscape of the last several and the next ninety-plus years, the one hundred years that define this "environmental century." No longer should we view the environment as something that divides us. No longer should we allow any faction of American society to claim that it alone is the guardian of the environment. No longer will innovators and outliers be described as "antienvironmental" without rebuking the charge. Our nation's motto, E *pluribus unum*, is a fitting platform for addressing environmental issues. As Americans, we must rally around an environmental game plan we can all support.

We hope that the many testimonials in this book will enable every citizen of our great nation to proclaim, "Of course I am concerned about the environment, isn't everybody?" It is time for each of us—whether our political views are right, left, or center—to stand up and demand an end to the squabbling. Instead, let's just agree to focus the talents of this great nation on working together to solve the real environmental problems confronting us. We are all members of a big-tent, bipartisan, multicultural "green team" whose commitment to the environment is genuine. Throughout this book we have advocated civil dialogue. It is rarely observed on television or radio anymore, but we will discover real solutions if we agree to talk about the issues frequently and in depth. Our presidential campaign season is a very poor example of dialogue and debate because it is comprised of photo opportunities, scripted sound bites, and pundit-laden spin sessions designed to keep candidates out of trouble on important national questions. We deserve better.

To arrive at consensus, we need to concentrate on areas of agreement rather than disagreement. This is the fundamental approach of Swedish physician Dr. Karl-Henrik Robert, who developed an environmental movement called The Natural Step, or TNS, as it is widely known. By issuing a draft of his ideas to fifty top scientists and revising it until all were satisfied with the final product, Robert decided to circulate the consensus document. By tenacious networking, he succeeded in distributing copies of the document in pamphlet and audio formats to every school and household in Sweden, reaching an audience of 8.7 million people. The principles derived from Robert's consensus provide that industry should phase out petroleum products and nonrecyclable materials, stop the use of nonbiodegradable compounds, ensure no net degradation of renewable capacity, and pass lessons on to developing nations. By advocating compliance as a long-term goal, Robert approaches company reform from a position of pragmatism. Compliance, in turn, drives innovation. TNS, therefore, is a compass that companies can apply to their business processes. TNS has already succeeded in transforming the practices of many big companies, including IKEA, one of the world's leading furniture manufacturers, Electrolux, and Swedish McDonald's. The Natural Step, advancing steadily and quietly, is an inspiring system of social and organizational change.

A lifelong commitment to any good cause is often inspired by legendary historical figures celebrated by our culture in poetry and song. On occasion, inspiration is found in less likely places, packaged, for example, in the personality of a bright girl with keen senses in Palm Beach, Florida. Ten-year-old Lilly Capehart has a special rapport with the many lizard species that flourish in this subtropical climate. The animals have responded by accepting her gentle touch. An entrepreneur at heart, and the daughter of a skilled photographer, she poses them, photographs them, and then markets the pictures as note cards and portraiture. Her love of nature and her business acumen have elevated the profile of these local creatures. Suddenly the Palm Beach lizards don't seem so lowly anymore. A mentor to peers and parents alike, Lilly is teaching her community how to live in joyful harmony with all living things. The lizards don't seem to suffer from their celebrity, and Lilly's commitment to their future expands with each backyard encounter, in their niche and on their turf. Kids such as Lilly someday grow up to be zoo directors, college professors, and even Speakers of the House.

Children comfortably interact with the fauna and flora of our living planet and experience close up the joy and the wonder of a bizarre and bountiful diversity of life under rocks, in the treetops, and below the surface of ponds, streams, rivers, and oceans. When children get close to nature, they are content to listen, touch, and smell life— and even taste of it, if they dare. A youthful Charles Darwin popped beetles into his mouth when his hands and pockets were overpopulated with the specimens he collected. He learned quickly that beetles taste bad, but he harbored the powerful zeal of a child enthralled with the natural world. His youthful enthusiasm for nature lasted a lifetime.

The multitude of mysterious small creatures in our world, caterpillars, for example, gladdens the idle child. Kids easily capture the slow-moving and often brightly colored caterpillars. They are also among the last terrestrial critters that are essentially unknown to science. We know there are many thousands in the United States alone, but most of them are unnamed.

Nature is a child's first great puzzle, and fascination with the natural world is a building block for science and for citizenship. To a large degree, our childhood experiences with nature compelled us to write this book. Inspired by a lifetime of close contact with nature, we were moved to commit our lives and our careers to protecting the denizens of this living earth. We believe Theodore Roosevelt aligned the priorities succinctly and spoke for us when he said:

The conservation of natural resources is the fundamental problem. Unless we solve that problem it will avail us little to solve all others.

Much like Roosevelt, who found inspiration in every nook and cranny of the known world, other explorers have found their inspiration at sites such as Crane Meadows in Nebraska. There, you can see some of the 500,000 sandhill cranes that crowd the Platte River system. During their occupation of the Platte River site, cranes significantly increase their body weight, as they prepare for the long migratory flight north where they will reproduce and sometimes face food shortages due to weather. Tourists and residents of Nebraska and the Midwest are indeed fortunate to experience and enjoy such a unique ecological treasure. However, Crane Meadows does not exist in a vacuum. It owes much of its success to Omaha-born entrepreneur Peter Kiewit, a generous philanthropist who died in 1979, but who left much of his fortune to his foundation. In April of 2000, the Kiewit Foundation announced a challenge gift of $550,000 to Crane Meadows, a generous legacy of facilities, amenities, and habitat developed for the entire nation to appreciate.

The whooping crane also visits Crane Meadows. It is a closely related but a rare cousin to the ubiquitous sandhill species. Whooping cranes were desperately in need of protection early in the twentieth century, but we came to their rescue—almost too late. By 1937, there were only *fifteen* whooping cranes left on Earth. Whatever we might think about the tendency of the federal government to become too involved in regulation, the whooping crane provides a good example of how the government plays a necessary and fundamental role in environmental protection. The realization that the crane population had dwindled to the brink of extinction prompted strict protection and monitoring by federal wildlife authorities. Slowly the population has grown to 130 free-ranging birds with another 40 living in a supplementary captive collection.

The power of the American government alone, however formidable, could not have saved the whooping crane. The list of individuals, organizations, and corporations that have assisted in the tenfold increase in the number of birds is far too long to be recounted here, but one source of support illuminates the breadth of collaboration. Orvis, the sporting outfitter, has donated tens of thousands of dollars to the International Crane Foundation, located in Baraboo, Wisconsin, to help increase the number of cranes through breeding programs. It may be a small victory to reach a level of 150 whooping cranes in the world, but we can take pride that conservation works. As we have learned, captive breeding can stave off extinction. Over time, cranes nurtured in zoos or in breeding sanctuaries can resurrect a dying population.

As compelling as any living crane, and a flagship of the conservation movement, is the giant panda. It is impossible not to love them. The panda's power to persuade led the World Wildlife Fund to use its image for the fund's inspirational logo. Their intriguing, infantile appearance —big eyes, enlarged head, and chubby extremities—are

characteristics biologists call *neotenic*. Recent studies by
Stephan Hamann, a neuroscientist at Emory University,
revealed that cute photographs, including giant pandas,
activate sites in the human brain associated with pleasure
and positive emotion. This innate response to neotenous
stimuli is thought to be tied to our need to nurture our
children (or anything that resembles them, for example,
the epitome of neoteny, our dogs and our cats). In essence,
our humanity is hardwired so that we cannot resist babies,
pets, and giant pandas. Cranes and pandas are both Asian
icons and are prominently featured in Chinese, Japanese,
and Indian art.

Our connection to nature runs deep; it's in our reli-
gions, our genes, our brain, and the soul of humanity.
Because we cannot deny it, there is good reason to nurture
and sustain it. Our powerful bond with the natural world
is the engine that drives our interest, commitment, and
action on behalf of the earth. The Contract with the Earth
is therefore entirely consistent with our human nature, our
beliefs, and our historical ties to great American environmentalists such as John Muir, Aldo Leopold, Henry
David Thoreau, and Theodore Roosevelt. Our affinity to
wildlife is especially powerful and animal imagery adorns
our automobiles (rabbit, sable, Taurus), our subdivisions
(Coyote Creek), our wine (Toad Hollow), and our local
sports teams (Bears, Lions, Marlins). Animals still inspire
and excite us just as they affected our ancient ancestors

who hunted and fished them, tamed them, worshiped them, and learned from them.

The potent personality of nature is surprisingly close at hand, and it's easier to access than we realize. Some of the best nature walks available can be found in cities, on trails crafted by groups dedicated to restoring nature in the urban landscape. "Naturalist's Walk" in New York City's Central Park is one of those special places. The Central Park Conservancy restored the site in 1996 to provide a rustic outdoor classroom within a naturalistic landscape. Native plants attract butterflies and birds and radiate colors of red, yellow, and purple. Azaleas and rhododendrons are featured with other woodland walking trails amid fields of ferns, snakeroot, yellow witch hazel, and fragrant black locust. Visitor amenities make it a pleasant venue for sitting and observing pond life. Further, the stunning return of beavers to New York City is an example of what can happen when we restore habitat to attract wildlife. Absent from New York since colonial times, the species was recently identified living free on the banks of the Bronx River just inside the boundaries of the Bronx Zoo. This discovery demonstrates to every urban dweller that recovery is feasible. Conservation works.

Once considered a dangerous city park, the Central Park campus has become a true oasis with a festive schedule of music and entertainment supplementing an experience in nature. Combining private money, marketing, and a contract with the City of New York, the user-friendly environment has never functioned better for people, wildlife, and living plants. Birders have identified 275 species of migratory birds in Central Park. Birds and people alike benefit from the park's 26,000 trees. Like the Central Park experiment, the organized greening of our cities can provide the restorative habitats that our urban citizens require and a visceral pathway to urban conservation. No less than a bluebird or a beaver, humans also crave natural habitats.

Especially evident in cities, our persistent estrangement from the natural world manifests a greater danger in its effects on children. Richard Louv has written an important book alerting parents to the deleterious consequences for children who lose contact with nature. In *Last Child in the Woods*, he asserts that children need exposure to nature for the healthy development of their senses, to learn and to create. Louv wrote:

> Nature—the sublime, the harsh, and the beautiful—offers something that the street or gated community or computer game cannot. Nature presents the young with something so much greater than they are; it offers an environment where they can easily contemplate infinity and eternity.

Apart from nature and restricted by hardscapes and electronic tethers, our children, according to Louv, run the risk of acquiring "nature deficit disorder," a malady that he describes as a contributing factor to a recognized mental health construct, attention deficit hyperactivity disorder, or ADHD. Nature deficit disorder, not yet acknowledged by scientists or clinicians, is a working hypothesis that Louv believes helps to explain the onset of ADHD, and he proposes that exposure to nature should be offered as a therapy for children who have difficulty attending to stimuli and learning in conventional classrooms. His call for research in this domain is compelling. Indeed, the one salient difference between the neighborhoods of our generation and the current environment we provide for our children is the diminution of nature. This is a trend that must be reversed.

At large in a technical world, we are refreshed each day by exposure to clean air and clear skies; we are disturbed when we encounter a polluted river, a reeking landfill, or smog. All of us, of every faith and at any intersection on the political spectrum, share this juxtaposition of awe and ire. Optimizing awe and eliminating ire is a worthy goal and a significant contribution to the nation's collective psychological well-being.

TALKING POINTS

1. Our nation's motto E pluribus unum reminds us that we can move proverbial mountains when we abandon adversarial politics and work together for a common cause.

2. Whooping cranes, giant pandas, and other charismatic species have inspired global conservation and stimulated unity and peaceful cooperation among many nations.

3. Our growing estrangement from nature is a dangerous trend, especially in view of its potential to damage the psyche of our children.

EPILOGUE

We need our reason to teach us today that we
are not . . . the lords of all we survey . . . we are the
Lord's creatures, the trustees of this planet, charged
today with preserving life itself . . .

MARGARET THATCHER

As we anxiously awaited the turn of a century and the benchmark of the millennium, the world was captivated by the doomsday vision of the dreaded Y2K phenomenon. Dire predictions drove computer upgrades in every household, business, and most nation-states throughout the world. Looking back at Y2K, it is easy to take the position that our vigilance worked because nothing bad happened when the clock struck midnight and we entered the twenty-first century. Somehow, our computers continued to crank out e-mail, and nobody lost his or her inventory of data. Y2K was barely a blip on the digital radar screen in 2000. One retrospective view of Y2K pertains to the estimated $300 billion cost of the systems overhaul as a bargain given the chaos and confusion that was likely averted. An unforeseen dividend was that some potential consequences of the 9/11 emergency were circumvented by technology installed for Y2K. Conversely, critics of the unprecedented remediation argue that the scale of the Y2K disruption was vastly overestimated so the huge expenditure of time and money may have been unnecessary. We do not know whether the mass hysteria associated with Y2K was a helpful or harmful reaction. However, since an early and orderly technology plan was executed worldwide, widespread fear and anxiety did not prevent business and government from proactive tinkering. In this case, rational implementation, according to the known facts, overcame collective hysteria.

On the scale of Y2K, Stanford professor Paul Ehrlich issued in 1968 an early catastrophic snapshot of a dismal human future in his best-selling book, *The Population Bomb*. Ehrlich predicted widespread world famine between 1970 and 1985 due to overpopulation and scarce resources. Instead, the world experienced a precipitous drop in fertility and food was exported by the United States at record levels because of the effects of the so-called green revolution in agriculture. The author has softened his tone in recent years, discussing his book in terms of "possibilities" rather than "predictions," and by clarifying his intent to prevent a worst-case scenario by alerting the world to a looming emergency. As we see it, the lesson of Ehrlich's publication is that media are positioned and primed to run with a horror story and they are unprepared to prevent hyperbole from running ahead of the science.

One person's hysteria may be regarded as a legitimate cry for help by another, but one thing is certain, no one has yet made a good living accurately predicting the future of the earth, including the world's best climate scientists. Given the uncertainty of current science and computer models, doomsday scenarios are not very helpful. However, media prompts succeed when they promote understanding by activating a rational problem-solving process. The complexity of the issue of global climate change is illustrated by the observations of the philosopher Michael Serres who advised:

> As of now we don't know how to estimate general transformations on such a scale of size and complexity. Above all, we surely don't know how to think about the relations of time and weather . . . For do we know a richer and more complete model of global change, of equilibria and their attractors, than that of climate and the atmosphere? We are trapped in a vicious circle.

To ensure unbiased reporting and misunderstanding, media professionals should fiercely protect their objectivity. Equally important, media accounts of data projections should be studiously cautious given the great potential for errors in measurement and the inherent limits of statistical sampling and inference. A scientist's comments hastily uttered at a press event frequently become the next day's sensational headline. A telling example is one expert's use of the phrase "highway to extinction" to describe the inevitable fallout from future climate trends. As journalists are trained to ask tough questions, they should be especially tough on doomsday theorists rather than milking a doomsday story for all the drama and angst it can produce.

Opinions and advocacy will always have a place in journalism, but they must be balanced by carefully researched and unbiased news stories. The need for media objectivity will be critical in the days, months, and years ahead as we debate the critical path and prioritize best practices to achieve a healthy environment. At the risk of promoting a commercial cliché, print and electronic media should strive to be fair and balanced. Adversarial, extremist monologues on either side of an issue or one-minute media sound bites will not produce good results for our nation. Instead, experts and political leaders need to break out of the box of managed, scripted, and highly charged point/counterpoint. A return to the subdued style and scholarly depth of the storied Lincoln-Douglas nineteenth-century political debates is a reform badly needed in our time. Media corporations should be working to achieve an unbiased, in-depth presentation of the original ideas, platforms, and philosophies of all political candidates so the American people can select their leaders based on trusted, reliable information sources.

In the foreword to this book, Professor Wilson revealed the etymology of the Latin *conservare*. It means, literally, to "keep, guard, observe." The bedrock principles of modern conservative thinking include the rule of law, traditional values, property rights, fiscal prudence, and patriotism. None of these concepts precludes a commitment to the environment, and all of them are compatible with entrepreneurial environmentalism, as we have defined it. Further, we believe these principles are supported by other political ideologies in America and certainly leave room for compromise in the give-and-take of our democratic system of government. Therefore, no one should be surprised that a conservative should be eager to express leadership on the environment. Likewise, it should not confound us when a conservationist acknowledges that conservative political principles are compatible with a new kind of environmentalism. Furthermore, as trained empiricists in history and psychology, respectively, we are eager to confirm that environmental policy should be consistent with the facts.

We recognize that global climate change is supported by a wealth of scientific data derived from a diversity of measurement techniques: ice borings, tree rings, water tables, ambient temperature trends, and the like. However, we still cannot be certain about the variance introduced by distinctly human activities. Should human behavior be a cause, to any extent, it wouldn't be surprising, given the role that human beings have played in other environmental events, for example, habitat destruction, deforestation, air and water pollution, and extinction. However, the debate about the origins and sources of climate change should not be left to scientists alone. As Princeton Geosciences Professor George Philander has written:

Earth's habitability is too important a matter to be left entirely to experts, especially when they contradict each other for reasons that are ideological rather than scientific. Everyone ought to participate in discussions of environmental policies and to that end should have a rudimentary understanding of the processes that make this a habitable planet.

In three recent reports from the National Academy of Sciences, climate scientists decry the lack of adequate systems for collecting, sharing, and modeling climate data. We must heed these calls and provide the scientific community with the resources they require to improve future climate projections. A key first step is the development of a sophisticated data-gathering system with appropriate investment in gathering and analyzing data. The maintenance and continuous improvement of a worldwide data system should be a high priority. Furthermore, the data should be available to everyone. Scientific debate and dissent should be encouraged in the pursuit of a thorough and comprehensive understanding of the complexities of our environmental systems.

Yet, how do we respond today in the face of current uncertainties? As we recognize the scientific evidence that the earth is experiencing a warming trend over the past 100 years and that this trend may have serious consequences for the future, we favor reducing carbon loading in the atmosphere as a bold forward step and a positive public value. The development of new technologies is the best way to achieve significant reductions in carbon loading. To maximize the rate at which we develop and diffuse these new technologies, we advocate significant public and private investment in basic research and a renewed commitment to math and science education at all levels. To these ends, we advocate the incentive of robust prizes to complement the current system of peer-reviewed grants and contracts. For example, prizes can be awarded for significant advances in carbon-reduction technology, energy-efficient motor vehicles, breakthroughs to a hydrogen-based economy, or innovations in superconducting magnetic transportation systems. We believe that research must be stimulated on both academic and industrial fronts throughout the nation. If we aggressively pursue these practices, we are liberated from a dependency on petroleum-rich dictatorships and positioned for success in the worldwide marketplace for cleaner, greener technologies.

We must never lose sight of the fact that reliable, affordable energy is indispensable to economic growth around the world and that economic growth is essential if we are fully committed to a healthier environment. In so many ways, in America and abroad, we can truly achieve "green through growth." Americans face the extraordinary challenge of bringing to bear science and technology, entrepreneurship, and the principles of effective markets, to enable all people to experience a green economy and green development and facilitate a higher quality of life for their families and communities.

In an important book, *The New Economy of Nature*, published in 2002, Gretchen Daily and Katherine Ellison concluded that

> the record shows that conservation cannot succeed by charity alone. It has a fighting chance, however, with well-designed appeal to self-interest. The challenge now is to change the rules of the game so as to produce new incentives for environmental protection, geared to both society's long-term well-being and individual self-interest.

It is essential that we continue to monitor the environment; sponsor research to learn as much as fast as we can, and develop new technology to reduce the output of greenhouse gases that may be contributing to the pace of global climate change. We should address the entire issue of climate change in a careful, rational way. If mitigation efforts should fail and climate changes begin to exert severe control over the environment, as *Newsweek* editorialist Fareed Zakaria suggested, we will need plenty of good will and political cohesion to make the tough decisions that "adaptation" will require. Worst-case scenario projections on the rise of sea levels over the next one hundred years are unreliable and imprecise, making a long-term, objective economic cost-benefit analysis extremely challenging. There is widespread agreement on one thing. We cannot afford to be wrong about global climate change. Soon, we will need to delineate a strategic action plan to deal with what we know.

Global climate change is prominent on the agenda of many captains of industry as demonstrated by the formation of working coalitions with leading environmental nongovernmental organizations, for example, the U.S. Climate Action Partnership comprising executives from Alcoa, BP, Caterpillar, DuPont, General Electric, and others. In addition, other national governments are also taking bold steps, including Australia where, by 2010, efficient compact fluorescent bulbs will replace incandescent lights throughout the nation. Although incandescent bulbs last 750 hours, fluorescent bulbs will burn for 10,000 hours. Energy efficiency is good for the earth regardless of the pace of climate change, and it can be achieved at relatively little cost in many cases.

The American government, however, continues to posture and vent, unable or unwilling to commit or act decisively. Government leadership on the environment is still an unresolved issue in America, but our nation's entrepreneurs and corporations have already committed significant human and financial resources to pursue aggressively carbon neutrality and other environmental best practices. With care and objectivity, the pace of innovation should be compatible with a growing and increasingly diversified economy. It is still early in the political season, but we see some encouraging signs that both Republicans and Democrats are warming to the idea of entrepreneurial environmentalism, an approach that should appeal to all political parties.

As we go to press with *A Contract with the Earth*, President George W. Bush has moved closer to acknowledging the significance of global climate change. In his 2007 State of the Union address, President Bush announced plans for an ambitious program of research and development for alternative fuels such as ethanol and a goal of reducing our nation's gasoline consumption by a factor of 15 percent in ten years. His brother, Jeb Bush, a former governor of Florida, may have gone further by announcing, at the end of his second term in 2006, the formation of an Interamerican Ethanol Commission to promote the use of ethanol throughout the Americas.

The momentum to develop ethanol and other forms of renewable energy is impressive, but all innovations seem to have a downside. After President Bush issued his proclamation on renewable energy sources, Steven McCormick, President and CEO of the Nature Conservancy, cautioned that a massive federal mandate for ethanol production could require the conversion of some 30 million acres of land, potentially compromising our nation's water and soil quality and precious wildlife habitat. In the production of ethanol, a massive crop of corn, sugar, or grasses is required, so manufacturers would have to be very careful to keep from victimizing forests or food sources in favor of new fuel products. Clearly, conservation innovation must go hand-in-hand with strategic economic development to reach its full potential. Close collaboration between industrial and environmental stakeholders is essential as new technologies are proposed and tested. Future presidents will surely find a way to vet their bold proposals with an appropriate subset of environmental and economic gurus so our national leaders are better prepared to deliver workable and effective environmental policies.

Religious leaders have also engaged the issue, and denominations from left to right, throughout the world, have taken advocacy positions on the environment. For example, Greenfaith, based in New Jersey, was founded in 1992 to "inspire, educate, and mobilize people of diverse spiritual backgrounds to deepen their relationship with the sacred in nature and restore the environment for future generations." A Greenfaith partnership with the Healthy Building Network (HBN) provides expertise to design and develop green building and maintenance guidelines for religious institutions. Similarly, the United Synagogue of Conservative Judaism established Green Sanctuaries to encourage green facilities and practices. Another noteworthy trend is the initiative to use only sustainably harvested palm fronds to celebrate Palm Sunday. In 2007, some 1,500 American churches are expected to distribute 364,000 "eco-palm stems." The churches participating in this program have agreed to pay a higher price for palms in order to contribute to a better environment in Mexico. Hundreds of other examples of religion-based environmentalism can be found on the Internet. The environment has become a moral and ethical issue for many religious groups, and this trend may produce a generation of environmental acolytes.

With the presidency and control of Congress at stake, liberal Democrats and conservative Republicans have much to fight about in the months ahead, but the urgency of the environment leaves room for common ground, cooperation, and compromise. We have but one Earth and our life, liberty, and pursuit of happiness depend on its vitality. Scholars and public officials have proclaimed, and we totally agree with the sentiment, that our children and grandchildren will rue the day we failed to take responsible action to clean up the earth. We have issued our call to cohesion and consensus at a time when the atmosphere is sizzling with the overheated rhetoric of the imminent national presidential campaign. In this context, we will continue to advocate the priority of the environment and the urgency of bipartisan unity. An entrepreneurial form of environmentalism is appropriate and timely for all Americans; it is good for our country, good for our economy, and good for the earth.

Given the anticipated extraordinary pace of scientific change in the next quarter century, a rapid and reliable strategy to achieve environmental protection is to empower entrepreneurial technological innovation coupled with the power of markets to shift resources to better outcomes with more choices of higher quality at lower cost. We strongly believe that entrepreneurial environmentalism is a superior approach to bureaucratic, litigious, unrestrained regulation, and we are betting there is strong public support for the former.

Because we are optimists, we are confident in humanity's ability to restore and to renew the natural world. In an active partnership with committed and creative entrepreneurial environmentalists just like you, we are committed to protecting the earth from all current and future threats. If you support the broad principles of this Contract with the Earth, we invite you to contribute your time and your ideas as we create together a new kind of environmental movement.

Finally, an enlightened epoch of mainstream environmentalism has arrived. We should take seriously our demanding role as twenty-first-century environmental stewards. It is our obligation and our opportunity.

SOURCES AND SUGGESTED READING

Among the numerous books and articles studied during the preparation of this book, the following sampling of sources is particularly recommended by the authors.

Aburto, G. The media and biodiversity conservation. In *Biodiversity Conservation in Costa Rica*, edited by G. W. Frankie, A. Mata, and S. B. Vinson, 257–265. Berkeley: University of California Press, 2004.

Adams, J. S., and McShane, T. O. *The Myth of Wild Africa: Conservation without Illusion.* Berkeley: University of California Press, 1996.

Baden, J. A. What we have learned since Earth Day, 1970. *Bridge News*, April 14, 1999.

Balling, J. D., and Falk, J. H. Development of visual preferences for natural environments. *Environment and Behavior* 14, no. 1 (1982): 5–28.

Barnes, P. *Who Owns the Sky? Our Common Assets and the Future of Capitalism.* Washington, DC: Island Press, 2001.

Bean, M. J. How Gingrich saved the Endangered Species Act. *Environmental Forum* 16, 1 (1999): 26–32.

Beattie, A., and Ehrlich, P. R. *Wild Solutions.* New Haven, CT: Yale University Press, 1999.

Brower, M., and Leon, W. *The Consumer's Guide to Effective Environmental Choices.* New York: Three Rivers Press, 1999.

Christensen, J. Unlikely partners create plan to save ocean habitat along with fishing. *New York Times*, August 8, 2006.

Clarke, T. W., Reading, R. P., and Clarke, A. L., eds. *Endangered Species Recovery: Finding the Lessons, Improving the Process.* Washington, DC: Island Press, 1994.

Collins, J. *Good to Great*. New York: HarperCollins, 2001.

Connelly, J. Philanthropy has helped environmental movement. *Seattle Post-Intelligencer*, April 22, 2000.

Conway, W. G. Zoo conservation and ethical paradoxes. In *Ethics on the Ark*, edited by B. G. Norton, M. Hutchins, E. F. Stevens, and T. L. Maple, 1–9. Washington, DC: Smithsonian Institution Press, 1995.

Daily, G. C., ed. *Nature's Services: Societal Dependence on Natural Ecosystems*. Washington, DC: Island Press, 1997.

Daily, G. C., and Ellison, K. *The New Economy of Nature: The Quest to Make Conservation Profitable*. Washington, DC: Island Press, 2002.

Daily, G. C., Soderqvist, T., Aniyar, S., Arrow, K., Dasgupta, P., Ehrlich, P. R., Folke, C., et al. The value of nature and the nature of value. *Science*, 2000, 289, 395–396.

Dalton, R. Doing conservation by numbers. *Nature* 442, no. 7098 (6 July 2006): 12.

Davison, R. P., Burger, W. P., Campa, H., III, Conry, P. J., Elowe, K. D., Frazer, G., Mason, D. C, Moore, D. E., III, and Nelson, R. D. *Practical Solutions to Improve the Effectiveness of the Endangered Species Act for Wildlife Conservation*. Wildlife Society Technical Review 05-1. Bethesda, MD: Wildlife Society, 2005, 1–14.

DeAlessi, M. *Saving Endangered Species Privately: A Case Study of Earth Sanctuaries, LTD*. Policy Study 313. San Francisco: Pacific Research Institute, August 2003.

Deutsch, C. H. Companies and critics try collaboration. *New York Times*, May 17, 2006.

Easterbrook, G. Global warming: Who loses—and who wins. *Atlantic Monthly*, April 2007, 299, 3, 52–66.

Easterbrook, G. *A Moment on the Earth*. New York: Penguin, 1995.

Ehrlich, P. R. *The Population Bomb*. New York: Sierra Club / Ballantine Books, 1968.

Esty, D. C., and Winston, A. S. *Green to Gold*. New Haven, CT: Yale University Press, 2006.

Fiesta-Bianchet, M., and Apollonio, M., eds. *Animal Behavior and Wildlife Conservation*. Washington, DC: Island Press, 2003.

Frank, A. A. Plug-in hybrid vehicles for a sustainable future. *American Scientist* 95, no. 2 (2007): 158–165.

Frankie, G. W., Mata, A., and Vinson, S. B., eds. *Biodiversity Conservation in Costa Rica*. Berkeley: University of California Press, 2004.

Fulton, K., and Blau, A. Cultivating change in philanthropy: A working paper on how to create a better future. Site hosted by Global Business Network and Monitor Company Group LLP. www.futureofphilanthropy.org.

Gardner, G. *Inspiring Progress: Religion's Contribution to Sustainable Development*. Washington, DC: Worldwatch Institute, 2006.

Gifford, R. Making a difference: Some ways environmental psychology has improved the world. In *Handbook of Environmental Psychology*, edited by R. B. Bechtel and A. Churchman, 323–334. New York: Wiley, 2002.

Gingrich, N. *Winning the Future: A 21st Century Contract with America*. Washington, DC: Regnery, 2005.

Green, K. Clouds of global-warming hysteria. *National Review Online*, May 8, 2006.

Hancocks, D. *A Different Nature: The Paradoxical Nature of Zoos and Their Uncertain Future*. Berkeley: University of California Press, 2001.

Hanson, E. *Animal Attractions: Nature on Display in American Zoos*. Princeton, NJ: Princeton University Press, 2002.

Hardin, G. The tragedy of the commons. *Science* 162, no. 1243 (1968): 1243–1248.

Hawkin, P., Lovins, A., and Lovins, H. *Natural Capitalism: Creating the Next Industrial Revolution*. Boston: Little, Brown, 1999.

Hershkowitz, A. *Bronx Ecology: Blueprint for a New Environmentalism*. Washington, DC: Island Press, 2002.

Hiss, T. *The Experience of Place*. New York: Knopf, 1990.

Hoage, R. J., ed. *Animal Extinctions: What Everyone Should Know*. Washington, DC: Smithsonian Institution Press, 1985.

Irvin, W. R. The Endangered Species Act: Prospects for reauthorization. In *Transactions of the Fifty-seventh North American Wildlife and Natural Resources Conference*, edited by R. E. McCabe, 642–647. Washington, DC: Wildlife Management Institute, 1992.

Jackson, R. J., and Kochtitzky, C. *Creating a Healthy Environment: The Impact of the Built Environment on Public Health*. Sprawl Watch Clearinghouse Monograph Series. Washington, DC, 1–20. www.sprawlwatch.org.

Jepson, P. Governance and accountability of environmental NGOs. *Environmental Science and Policy*, 2005, 8, 515–524.

Karesh, W. B., and Cook, R. A. The next pandemic? The human-animal link. *Foreign Affairs*, July–August 2005, 38–50.

Kellert, S. R. *The Value of Life: Biological Diversity and Human Society*. Washington, DC: Island Press, 1996.

Kellert, S. R., and Wilson, E. O., eds. *The Biophilia Hypothesis*. Washington, DC: Island Press, 1993.

Kerry, J., and Kerry, T. H. *This Moment on Earth: Today's New Environmentalists and Their Vision for the Future*. New York: Public Affairs, 2007.

Kotler, P., and Roberto, E. L. *Social Marketing*. New York: Free Press, 1989.

Kristof, N. D. Another small step for earth. *New York Times*, July 30, 2006.

Lackey, R. T. Defending reality. *Fisheries*, 2001, 26, 29.

Lackey, R. T. Science, scientists, and policy advocacy. *Conservation Biology* 21, no. 1 (2007): 12–17.

Ladle, R., Jepson, P., and Whittaker, R. J. Scientists and the media: The struggle for legitimacy in climate change and conservation science. *Interdisciplinary Science Reviews*, 2005, 30, 3, 231–240.

Leopold, A. *A Sand County Almanac*. New York: Oxford University Press, 1949.

Lichter, S. R., and Rothman, S. *Environmental Cancer—A Political Disease?* New Haven, CT: Yale University Press, 1999.

Lieberman, B. Don't rush to judgment on U.N.: IPCC global warming summary. WebMemo, 1351. February 7, 2007, Washington, DC: Heritage Foundation.

Liftin, K. T., ed. *The Greening of Sovereignty in World Politics*. Cambridge, MA: MIT Press, 1998.

Louv, R. *Last Child in the Woods: Saving Our Children from Nature-Deficit Syndrome*. Chapel Hill, NC: Algonquin Books, 2005.

Mittermeier, R., Gil, P. R., Hoffman, M., Pilgrim, J., Brooks, T., Mittermeier, C. G., Lamoreux, J., and da Fonseca, G. A. B. *Hotspots Revisited*. Mexico City: CEMEX, 2002.

Moss, C. *Elephant Memories*. Chicago: University of Chicago Press, 1988.

Nattrass, B., and Altomare, M. *The Natural Step for Business: Wealth, Ecology, and the Evolutionary Corporation*. Gabriola Island, BC: New Society, 2002.

Neumayer, E. Do democracies exhibit stronger international environmental commitment? A cross-country analysis. *Journal of Peace Research* 39, no. 2 (2002): 139–164.

Newton, J. L. *Aldo Leopold's Odyssey.* Washington, DC: Island Press, 2006.

Norton, B. G. *Toward Unity among Environmentalists.* New York: Oxford University Press, 1991.

Norton, B. G., Hutchins, M., Stevens, E. F., and Maple, T. L., eds. *Ethics on the Ark: Zoos, Animal Welfare, and Wildlife Conservation.* Washington, DC: Smithsonian Institution Press, 1995.

Ostrom, E., Burger, J., Field, C. B., Norgaard, R. B., and Policansky, D. Revisiting the commons: Local lessons, global challenges. *Science* 284, no. 5412 (1999): 278–282.

Pahl, G. *Biodiesel: Growing a New Energy Economy.* White River Junction, VT: Chelsea Green, 2005.

Peng, C., Ouyang, H., Gan, Q., Jiang, Y., Zhang, F., Li, J., and Quiang, Y. Building a "green" railway in China. *Science*, 316 (2007, April 27): 546–547.

Pielke, R., Prins, G., Rayner, S., and Sarewitz, D. Lifting the taboo on adaptation. *Nature* 445, no. 8 (February 2007): 597–598.

Raven, P. H., ed. *Nature and Human Society: The Quest for a Sustainable World.* Washington, DC: National Academy Press, 1997.

Robert, K.-H. *The Natural Step Story: Seeding a Quiet Revolution.* Gabriola Island, BC: New Society, 2002.

Roman, J. A whooping success: the world's most endangered crane makes a comeback. *Wildlife Conservation*, June 2007, 38–45.

Rutherford, F. J., and Ahlgren, A. *Science for All Americans.* New York: Oxford University Press, 1989.

Sabini, M., ed. *The Earth Has a Soul: The Nature Writings of C. J. Jung*. Berkeley, CA: North Atlantic Books, 2002.

Scarlett, L. Moving beyond conflict: private stewardship and conservation partnerships. Heritage Foundation Lecture No. 762, September 27, 2002.

Serres, M. *The Natural Contract*. Ann Arbor: University of Michigan Press, 1995.

Sommer, R. *Tight Spaces: Hard Architecture and How to Humanize It*. Englewood Cliffs, NJ: Prentice-Hall, 1974.

Spotila, J. R., and Paladino, F. V. Parque Marino Las Baulas: Conservation lessons from a new national park and from 45 years of conservation of sea turtles in Costa Rica. In *Biodiversity Conservation in Costa Rica*, edited by G. W. Frankie et al., 194–209. Berkeley: University of California Press, 2004.

Stafford, N. Gas for the greenhouse. *Nature*, 442 (3 August 2006): 499.

Staudt, A. Huddleston, N., and Rudenstein, S. *Understanding and Responding to Climate Change*. Washington, DC: National Academy of Sciences, 2006, 1–24.

Tal, A., ed. *Speaking of Earth: Environmental Speeches That Moved the World*. New Brunswick, NJ: Rutgers University Press, 2006.

Terborgh, J. *Requiem for Nature*. Washington, DC: Island Press, 1999.

Van der Ryn, S., and Cowan, S. *Ecological Design*. Washington, DC: Island Press, 1996.

Vitousek, P. M., Mooney, J. A., Lubchenco, J., and Melillo, J. M. Human domination of earth's ecosystems. *Science*, 277 (1997): 494–499.

Wilcove, D. S. *The Condor's Shadow: The Loss and Recovery of Wildlife in America*. New York: W. H. Freeman, 1994.

Wilson, E. O. *The Creation: An Appeal to Save Life on Earth.* New York: W. W. Norton, 2006.

Wilson, E. O. The future of life—preserving our natural capital. *AEI Newsletter,* May 1, 2001.

Wilson, E. O. *The Future of Life.* New York: Knopf, 2002.

Wilson, E. O. *In Search of Nature.* Washington, DC: Island Press, 1996.

Yi-Fu, T. *Topophilia.* New York: Columbia University Press, 1974.

Zakaria, F. Global warming: Get used to it. *Newsweek,* February 19, 2007, 43.

OTHER USEFUL LINKS

The National Arbor Day Foundation: www.arborday.org

American Solar Energy Society: www.ases.org

Association for Zoos and Aquariums: www.aza.org

Center for Conservation and Behavior: www.centerforconservationandbehavior.org

Conservation International: www.conservation.org

The Conservation Fund: www.conservationfund.org

Earth Biofuels, Inc.: www.earthbiofuels.net

EarthEcho International: www.earthecho.org

Environmental Evaluation & Cost-Benefit News: www.envirovaluation.org

Foundation for Research on Economics and the Environment: www.free-eco.org

Geoplasma, LLC: www.geoplasma.com

Dian Fossey Gorilla Fund International: www.gorillafund.org

Green Belt Movement: www.greenbeltmovement.org

The Green Institute: www.greeninstitute.org
Green Strategies, Inc.: www.greenstrategies.com
The Izaak Walton League of America: www.iwla.org
National Fish and Wildlife Foundation: www.nfwf.org
The Natural Step Foundation: www.naturalstep.org
The Nature Conservancy: www.nature.org
New Environmentalism: www.newenvironmentalism.org
Newt Gingrich: www.newt.org
Natural Resources Defense Council: www.nrdc.org
National Wildlife Federation: www.nwf.org
Republicans for Environmental Protection:
    www.repamerica.org
Sustainable Business Institute:
    www.SustainableBusiness.com
Wangari Maathai: www.wangarimaathai.or.ke
WaterWebster: www.WaterWebster.org
Wildlife Conservation Society: www.wcs.org
The Wildlife Society: www.wildlife.org
The World Watch Institute: www.worldwatch.org

INDEX

Gates, Bill and Melinda, 119
Gelinas, Johanne, 168
General Motors, 97
geoplasma, 89, 90
Georgia, 110; colleges, 21
Georgia Department of Natural Resources, 111
Georgia Power, 98
Georgia Tech, 90, 161
giant panda, 178
Gingrich, Newt, 111
global climate change, 39, 113, 119, 142, 187, 189, 192, 194
Goldman Sachs, 117
Good to Great, 156
Google, 119
Google Foundation, 119
Gore, Al, 148
gorillas, 27
government, 13
Grand Canyon National Monument, 110
Grand Cayman, 80
gray wolf, 53
Great Lakes, 133
Green Belt Movement, 136
green development, 6
green energy, 88, 98
green enterprise, 5, 12
Greenfaith, 195

greenhouse gas emissions, 34, 64, 78, 92, 94, 96
green partnerships, 69
green revolution, 51, 186
Green River, 68
Green River Dam, 69
greens, 168
green sanctuaries, 195
green strategic thinking, 159
Green to Gold, 159
Gregoire, Christine, 46
Gresham, Oregon, 144
ground-level ozone, 135
Gulf Coast, 104
Gulf Stream, 99

habitat destruction, 189
habitat protection, 46, 166
Hamann, Stephan, 179
hard architecture, 128
Hardin, Garrett, 23
hardscapes, 181
Harvard, 119
Hayes, Denis, 118
Healthy Building Network, 195
Hereford, Texas, 90
Hewlett-Packard, 91
higher education, 12
highway to extinction, 187
Hippocrates, 37
Hoffmeister, John, 139

NEWT GINGRICH was Speaker of the U.S. House of Representatives from 1995 to 1999 and is widely heralded as the chief architect of the Republican Contract with America. Since his days as a college professor in western Georgia, where, in the early 1970s, he was an environmental studies professor, he has been involved in a variety of environmental initiatives. He was the founding chair of the West Georgia College Chapter of the Georgia Conservancy. He has championed various environmental causes, including efforts to create the Chattahoochee River Greenway, protect the wild tigers of Asia, and establish the Northwestern Hawaiian Islands National Marine Sanctuary.

TERRY L. MAPLE is president and CEO of the Palm Beach Zoo and professor of conservation and behavior at the Georgia Institute of Technology. Dr. Maple is a former president of the Association of Zoos and Aquariums, a fellow of the American Psychological Association, the founding editor of *Zoo Biology*, and a co-editor of *Ethics on the Ark* (Smithsonian Institution Press). He was president and CEO of Zoo Atlanta from 1985 to 2003.

NEWT AND TERRY ON

CONSERVATION AND THE ENVIRONMENT

"There's no question acid rain is a problem. Industry has lost that battle and is foolish to continue to fight it."
—Newt Gingrich, *Fortune Magazine*, 1988

"Our conservation stories must give our patrons hope and provide opportunities to contribute to projects and programs that can be successful. We can do this because the zoo itself is an example of success. Our living collection is an inspiration to those who would help us to save the wild, and in this way the animals are truly ambassadors for their own kind."
—Terry L. Maple, *ZooMan*, Longstreet Press, 1993

"The psychological well-being of humankind is clearly elevated by the presence of a multitude of living things, and we are diminished by the loss of other species . . . It is up to us to assert boldly the integrity of the whole and to preserve nature's magnificent diversity."
—Terry L. Maple, *Defenders Magazine*, 1995

"I believe that we must look at the entire ecosystem when dealing with the protection of endangered species."
—Newt Gingrich, *The Oregonian*, 1995

"We need to look at incentives for saving species and incentives for biodiversity . . . as opposed to a command and control model of punishing people, fining people, regulating people, and threatening people with jail, which I think, in a free society . . . in the long run [is] not as effective a way to change behavior as incentives."

—Newt Gingrich, *Reason* Magazine, 1996

"Frankly, every species we lose is a level of knowledge that is irreplaceable."

—Newt Gingrich, *San Jose Mercury News*, 1996

"Florida has the opportunity to become a laboratory that the entire world studies . . . There are very few places where you have a complex fragile ecosystem this close to this many people."

—Newt Gingrich, Associated Press, 1997

"Africa's elephants are huge and visible creatures. Their decline throughout their range has been swift and dramatic. They are a symbol and a metaphor for our protection efforts . . . poachers continue to annihilate whole populations of elephants throughout Africa . . . a fresh kill by poachers is a deeply disturbing event. It is an image that is lasting and something you cannot forget."

—Terry L. Maple, *Congressional Record*, 1997

"The highest investment priority in Washington should be to double the federal budget for scientific research. No other federal expenditure would create more jobs and

wealth or do more to strengthen our world leadership, protect the environment, and promote better health and education for all Americans. For the security of our future, we must make this investment now."
—Newt Gingrich, *The Washington Post*, 1999

"Doomsday messages just don't work anymore, if they ever did . . . research has demonstrated that scare tactics are not effective devices in teaching . . . far better, I think, to give your supporters reason to hope."
—Terry L. Maple, *Saving the Giant Panda*, Longstreet Press, 2000

"For much of this country's history, prizes motivated sharp minds to innovate quickly while avoiding the dual demons of massive paperwork and entangling bureau-cracies. Today, when the country needs breakthrough solutions in a wide range of pressing issues—among them, health care, the environment, security and space—prizes could serve the U.S. government well."
—Newt Gingrich, *USA Today*, 2002

"It is possible to have a healthy environment and a healthy economy. It is possible to build incentives for a cleaner future. It is possible to have biodiversity and wealthy human beings on the same planet. And it is possible to have free markets, scientific and technological advances, and an even more positive environmental outcome. There is every reason to be optimistic that if we develop smart

environmental and biodiversity policies our children and grandchildren will experience an even more pleasant world."
—Newt Gingrich, *Winning the Future*, Regnery, 2005

"Future zoos and aquariums must continue to function as ethical, caring, full-service habitats for rare and endangered wildlife. Their patrons will no doubt require justification for constraining and confining such an array of complex creatures, and our zoos and aquariums must be therefore tireless advocates for their wild kin and the remaining protected ecosystems of the world."
—Terry L. Maple, *Museum Philosophy for the Twenty-first Century*, Alta Mira Press, 2006

"One area of reform that Republicans and Democrats alike could learn from was [Teddy Roosevelt's] approach to the environment. He understood that conservative and conservation have the same root and he was passionately committed to conserving America's natural resources for future generations . . . Republicans would do well to study his commitment to national parks, national forests . . . and the natural world . . . Democrats would do just as well to note that Theodore Roosevelt saw man as part of nature and not as its opponent."
—Newt Gingrich, *Time Magazine*, 2006

The authors have pledged a portion of the royalties from *A Contract with the Earth* to support the Tigers Forever Asian field program of the Wildlife Conservation Society of New York.